THE ORIGINS OF PTOLEMY'S ASTRONOMICAL PARAMETERS

by

Robert R. Newton

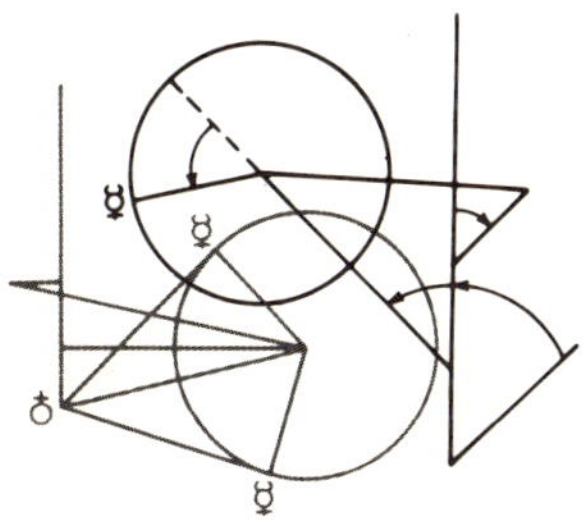

Technical Publication No. 4

A cooperative publication by

The Center for Archaeoastronomy, University of Maryland

The Johns Hopkins University, Applied Physics Laboratory

The Technical Publications of the Center for Archaeoastronomy:

1. *Archaeoastronomy in the Americas: Abstracts of the June 1979 Santa Fe Conference,* Ray A. Williamson, ed., 1979

2. *Archaeoastronomy in the Americas*, Ray A. Williamson, ed., 1981. A Center for Archaeoastronomy/Ballena Press cooperative publication

3. *When the Stars Came Down to Earth: Cosmology of the Skidi Pawnee Indians of North America*, Von Del Chamberlain, 1982. A Center for Archaeoastronomy/Ballena Press cooperative publication

The Books of Robert R. Newton:

Ancient Astronomical Observations and the Accelerations of The Earth and Moon, 1970

Medieval Chronicles and the Rotation of the Earth, 1972

Ancient Planetary Observations and the Validity of Ephemeris Time, 1976

The Crime of Claudius Ptolemy, 1977

The Moon's Acceleration and Its Physical Origins: Volume 1, As Deduced from Solar Eclipses, 1979

ISBN 0-912025-02-6

The Center for Archaeoastronomy is an independent institute organized within the University of Maryland and associated with the Astronomy Program and the Department of Anthropology. Housing is provided by the Mathematical and Physical Sciences and Engineering Division.

The Applied Physics Laboratory is a division of The Johns Hopkins University and operates under contract No. N00024-83-C-5301 with the Department of the Navy.

CONTENTS

FIGURES

TABLES

PREFACE

More than ten years ago I began a program of studying the non-gravitational forces in the solar system, and of studying their time dependence, by analyzing astronomical measurements made in ancient and medieval times. Early in the course of this study I was struck by the fact that all the observations I analyzed gave a cohesive picture except for the observations that Claudius Ptolemy claimed to have made himself in his famous treatise on astronomy, and I spent considerable effort in trying to find out why his observations are so discordant with all others.

My first approach was to think of methods of observation that might have yielded the large errors we find in Ptolemy's data, but I had no success. Then one day it occurred to me to see if his solar data had been fabricated, and I soon discovered that they were. Earlier writers had also speculated that this was the case, and J. B. J. Delambre in 1819 and J. P. Britton in 1967 had actually and independently published proofs that some of the solar data were fabricated. However, their writing on the subject had been totally ignored in the literature, which explains why Britton could independently make a discovery that had been published almost 150 years before his own work, and why I could later make the same discovery independently of both of them.

In *The Crime of Claudius Ptolemy* (Johns Hopkins University Press, Baltimore and London, 1977) I analyzed all the observations reported in Ptolemy's treatise. I found that all of Ptolemy's own "observations" that he used and that can be tested proved to be fabricated. Amazingly, a large fraction of the observations that he attributes to other astronomers were also fabricated. I am still flabbergasted that he would dare to do this.

The realization that Ptolemy fabricated most of his data forces us to reexamine much of what we have thought that we know about Greek astronomy. Much new research into Greek astronomy will be needed before we can straighten out this situation, and the present work is only a beginning. The main goal of this work is connected with the following problem: When Ptolemy fabricated his data, he did so using his astronomical theories, which contain certain numerical parameters. While his theories contain errors that are large compared with the errors in genuine observations made by Greek astronomers, they still conform approximately to reality. Manifestly, he could not have derived his parameters from his own "observations", since the observations were in fact derived from the parameters. Whence, then, did he get his parameters?

In the list of references that appears near the end of this work, I include only those works that I have directly consulted. In some cases I have thought that a particular matter was not sufficiently relevant for me to locate and study the original sources, and I have relied on secondary sources. In these cases, I give the primary sources only parenthetically or in footnotes.

In citing a reference, I use two items. The first is the name of the author in italics and the second is the year of writing or publication. If a work is issued by an organization rather than an individual, I use part of the title or part of the name of the organization as the first item. The second item is always in square brackets, and the first may or may not be, depending upon the context. For example, I cite the works by Delambre and Britton that have been mentioned as *Delambre* [1819] and *Britton* [1967]. If it is necessary to identify a specific place within a reference, such as a page or a section, that information appears after the year within the brackets. If the exact year is uncertain, I give a year for the sake of definiteness in citation but I precede the year by the symbol "ca.". If a particular work is cited frequently within a short space, I often omit the date and use the author's name without italics.

I thank Dennis Rawlins and Owen Gingerich for sending me advance copies of the papers that are cited under their names. I also thank my colleagues B. B. Holland and R. E. Jenkins for a critical reading of this work and for many valuable discussions. I particularly thank Lynn Hoff and Sheila Westbrook, of The Applied Physics Laboratory, for their assistance in preparing this work for the press. This work, which is intimately connected with studies in the precise measurement of time, was supported by the Department of the Navy under contract with The Johns Hopkins University. Much of the material in Sections I.1 and I.2 was used in an article about astrology that appeared in the *Johns Hopkins APL Technical Digest,* Volume 3, Number 1, 1982.

CHAPTER I

BACKGROUND AND PURPOSES OF THIS WORK

1. Astronomy and Astrology

Once only in my life, in a remote village in the west of England, I had the opportunity of being out of doors on a cloudless night where I could see no evidence of artificial light. The sight was breathtaking, but it was a sight that most people could see every clear night until a relatively short time ago. It is little wonder that man has studied the stars since time immemorial.

Early man, I am sure, quickly learned to distinguish the sun and moon from the other objects in the heavens, partly because of their brightness and partly because of their relatively large apparent size, and the motions of the sun and moon have been involved in all the calendars that I know about. People needed calendars in order to know when to go hunting, when to plant their crops, and when to fight their wars. The need for a calendar may have been one of the original forces that led people to study the stars, that is, to study astronomy.

All the other objects in the heavens (except for a few transitory objects like comets and meteors) are effectively points of light, and I do not know when people first realized that five of these points move around with respect to the thousand or so others that can be seen with the unaided eye, and that seem to remain fixed. The motions of these five points, which we call the planets,[1] are rather complicated. Most of the time they travel from west to east with respect to the fixed stars, but occasionally they stop, travel backward for a while, and then stop again before resuming their normal courses. Can we blame early people who, observing these peculiar and majestic movements of the planets, endowed them with the attributes of gods?

[1]A word that is derived from a Greek word meaning "wanderer."

Some one, some where, at some time developed an idea that, though it is ridiculous, has had an important and probably deleterious effect on the lives of millions or even billions of people. This is the idea that the positions of the planets at the instant of a person's birth influence or even control the course of his life. The pseudo-science that we call judicial astrology is the embodiment of this idea.

So far as I know, this idea must be charged against the Greeks and Babylonians, among whom it appeared almost simultaneously. To be sure, earlier civilizations had a well-developed set of what we may call *omina*. The *omina* were a set of appearances in the heavens which may have predicted the future according to some ideas, or may have shown the attitude of the gods according to other ideas. But the *omina* did not have the force that the positions of the planets assumed in judicial astrology; they merely warned a person, usually a monarch. Here we have the positions of the planets at a specific instant[2] assuming a dominant influence over a private person's entire lifetime.

In the Greek civilization, and also in European civilization until about the time of the Renaissance, little distinction was made between astrology and astronomy. Where a distinction was made, we can see in the very names which subject was considered the more important. Astrology means the science of the stars whereas astronomy merely means their arrangement.

The reason for this hierarchy is not hard to see. When a child was about to be born into a royal household, its parents were perhaps able to retain a corps of observers to keep watch over the heavens and to record the positions of the planets at the instant of birth, if the sky was clear. But most families could not afford this luxury, so most astrology had to be based upon the calculation of the planetary positions at an arbitrary time for which there were no observations. Thus astrologers had to try to find the laws that governed the motions of the planetary stars. As *Wells* [1922, p. 731] says, the astrologers "were ... a 'practical'[3] race. They studied the stars — to tell fortunes. They lacked that broader faith and understanding which induced men simply to study the stars."

Thus we have seen two reasons why "practical" people might have studied the stars and begun the development of what we now call astronomy. One was to develop a calendar and the other was to supply the celestial positions needed for judicial astrology. As I said a moment ago, little distinction was made between astrology and astronomy in western society until roughly the time of the Renaissance, and an astronomer was expected to be an astrologer as well. We may imagine

[2]In some systems of judicial astrology, the positions of the signs of the zodiac also had to be considered.

[3]The quotation marks are in the original.

that many astronomers were happy with this arrangement, but I suspect that others were not. Besides the "practical" people, I suspect that there have always been people who studied a subject such as astronomy purely out of curiosity about it. That is, there were people who did have the faith "simply to study the stars."

2. Claudius Ptolemy

We do not know much about Ptolemy's life. His first name seems to be Greek and his second name Egyptian. There is no reason to connect him with the royal line of Ptolemys who were at least the nominal rulers of Egypt for some time; he may have carried the name Ptolemy because he came from one of the towns in Egypt that were named after the royal line. However, all statements to this effect are late and unsupported by any real evidence.

There is a standard "portrait" of Ptolemy that is sometimes printed, and there is also a standard description of his physical appearance. I think it goes without saying that these are late creations with no validity.

Ptolemy wrote extensively on many subjects, but the work for which he is most famous today is *'E Mathematike Syntaxis* [*Ptolemy*, ca. 142], which I shall call simply the *Syntaxis* in the rest of this work. This is a lengthy treatise on astronomy that is divided into thirteen books, each of which is divided into parts that are called chapters in most modern writing on Ptolemy. I shall cite a particular chapter by writing, for example, Chapter III.7. By this I mean the seventh chapter in Book III.

In the *Syntaxis* Ptolemy refers to the emperors Hadrian (emperor from 117 to 138) and Antoninus Pius (emperor from 138 to 161). In addition he claims to have made certain astronomical observations with dates ranging from 127 March 26 to 141 February 2. I have shown [*Newton*, 1977][4] that all these observations were fabricated, so the dates themselves do not supply a rigorous dating of Ptolemy's activities. Still, he must have wanted the *Syntaxis* to be read as soon as possible after he finished it, and thus he would presumably have used dates that would be plausible. That is, the dates furnish a rough indication of the time when he wrote the *Syntaxis*, even though the observations with those dates were fabricated. To say that he finished it about 142 is a reasonable guess.

[4] I shall designate this reference as *CCP* in the rest of this work. The letters are the initial letters of the principal words in the title.

Some later writers have claimed that Ptolemy lived into the reign of Marcus Aurelius, who was emperor from 161 to 180. There is nothing implausible in this, but there is no evidence that it is correct.

Ptolemy says in several places that he made his observations in Alexandria, and this is the only place that he mentions in respect to himself. However, some manuscripts of his works have an insertion which says that an inscription was dedicated to him in the town of Canopus in the tenth year of Antoninus Pius, and thus in about 148. The inscription presumably contained lists of astronomical parameters taken from the *Syntaxis*.

Canopus, whose modern name is Abukir,[5] is a seaport that is much older than Alexandria and that lies about 20 kilometers to the northeast. It was the base for a Roman fleet in imperial times, and it was renowned in the Roman Empire for the dissolute and licentious nature of life there.

On the basis of the manuscripts which refer to the "Canopic inscription", some historians have concluded that Ptolemy lived and worked in Canopus rather than Alexandria. Opinions about the Canopic inscription are varied. *Neugebauer* [1957, pp. 195 and 207] explicitly takes Ptolemy to be the author of the Canopic inscription, but without drawing any conclusions about his place of residence. *Pedersen* [1974, p. 12] refers to the "undoubted tradition" that Ptolemy erected an inscription at Canopus, and concludes that Ptolemy had some sort of observatory in Alexandria but that he probably did his work, other than making observations, in Canopus.[6] *Toomer* [1975, p. 186] doubts that the Canopic inscription was authentic, and he sees no reason to suppose that Ptolemy ever lived or worked anywhere other than Alexandria (in his adult life, that is; he might have been born elsewhere). Toomer's position seems sound to me.

Ptolemy wrote many things besides the *Syntaxis*. Sometime after the *Syntaxis* he wrote what we may call a popularized version of it called *Planetary Hypotheses* (but see the further discussion of this work in Section IV.6). He also prepared a work that is called *Handy Tables*, which is a collection of tables, with instructions on how to use them, from which one can find astronomical positions by simple calculations. This is basically a collection into one place of the various tables that are scattered throughout the *Syntaxis*, but with some changes in methodology

[5]There is an interesting agreement of names here. Abukir means "Father Kir", and Kir is a form of the name Cyrus or Syrus. Ptolemy addressed many of his writings, including the *Syntaxis*, to an unknown person named Syrus. However, the "Kir" for whom Abukir is named is believed to have been a saint of the Coptic church. If so, there is no significance in the agreement of the names.

[6]Pedersen makes an amusing but minor slip here. He refers to Canopus as a town "... which offered better possibilities for a quiet life of study ..." than "noisy" Alexandria. This about a city that was famous in the Roman world for its dissoluteness.

[*Pedersen*, 1974, pp. 397ff], and with lower precision. His other work on astronomy is called *Phases of the Fixed Stars*, in which Ptolemy gives tables from which one can find the dates of the first and last visibility of various important stars in the morning and evening, for several latitudes, along with weather predictions[7] based upon these "heliacal" risings and settings. All these works still survive, but the *Handy Tables* exists only in a revision made by Theon of Alexandria about 360.[8]

Other surviving works include treatises on optics, harmony, and geography; the geography in particular was a famous work. There are also some minor surviving works, and some works that are mentioned by ancient writers but that have been lost.

There is one other important work that still survives. The title of this work means something like *Astrological Influences*, but it is commonly known even in English by the name *Tetrabiblos*, from the fact that it is divided into four books. I have not read the *Tetrabiblos* myself, and the following description is taken from *Toomer* [1975, p. 198]. In the opening chapters of Book I, Ptolemy defends astrology against the charge that it cannot accomplish what it claims. Among his arguments is one which says that, since the sun and moon have obvious physical effects, all other heavenly bodies must also have significant physical effects. Toomer uses the adjective "ludicrous" to describe Ptolemy's arguments on this point.[9] After the defense of astrology, the rest of the first book explains the technical concepts of astrology. Book II deals with the effects of astrology on the earth in general, such as weather prediction, and the last two books deal with astrological influences on human beings. The *Tetrabiblos*, like the *Syntaxis*, the *Planetary Hypotheses*, and the *Handy Tables*, is addressed to the unknown Syrus.[10]

Pedersen [1974, p. 400] says of the *Handy Tables*: "There can be no doubt that the tables were composed for the benefit of people who had to perform astronomical calculations in great numbers, but without

[7] *Pedersen* [1974, p. 406] calls these "astrological weather predictions". I have not read the work in question, but it seems to me that this judgment may not be justified. For example, the "heliacal rising" of Sirius occurs in July. It is not astrological to predict that this event will be followed by hot, sultry weather in Washington, D. C. In fact, this event is the origin of the term "dog days", since Sirius (α Canis Majoris) is also known as the Dog Star. The heliacal risings and settings furnished the easiest way to keep up with the seasons in early societies.

[8] However, *Neugebauer* [1957, p. 198] says that Theon was the author of the *Handy Tables*. See footnote #10.

[9] *Pedersen* [1974, p. 402], on the other hand, describes them as "an intelligent piece of reasoning".

[10] I have not consulted the *Handy Tables* myself, and I am relying on *Toomer* [1975, p. 187] and *Pedersen* [1974, p. 397] for the information that the tables are addressed to Syrus. If this is correct, it seems to me that this increases the probability that they are due to Ptolemy himself and not to Theon. See footnote #8.

any high accuracy. Only one class of scientists fulfills these conditions, namely the astrologers."

Until astrology began to lose part of its grip on educated Europeans, say about 1600,[11] Ptolemy was more famous for the *Tetrabiblos* than for the *Syntaxis*. Of these two works, the *Tetrabiblos* was the first to be translated into Arabic, into medieval Latin,[12] and into English, and it was also the first to be printed [*Pedersen*, 1974, pp. 14ff and p. 400].

Most modern writing about Ptolemy emphasizes the *Syntaxis* and his treatise on geography, while giving only passing mention to his writing about astrology. I think it is time to redress this balance. Ptolemy does not sound to me like an astronomer who is forced by the pressure of the times to cast an occasional horoscope. Instead he sounds like a strong advocate of astrology. We shall even have to ask what his primary motivation was. Was he primarily an astrologer, who was interested in astronomy only because of the need to calculate celestial positions? Or was he primarily an astronomer, but one who happened to believe in astrology? Or was he perhaps both? I suspect that the latter is the answer and I shall return to this topic in Section IX.8.

Gingerich [1980], in the second sentence of his paper, gives the following quotation: "I know that I am mortal by nature, and ephemeral; but when I trace at my pleasure the windings to and fro of the heavenly bodies I no longer touch earth with my feet: I stand in the presence of Zeus himself and take my fill of ambrosia, food of the gods." He notes that some of the manuscripts of the *Syntaxis* open with this quotation, and he then goes on to say: "The epigram seems to place Ptolemy within the long series of scientists who have tasted the intoxicating pleasure of a splendid theory."

I am afraid that this epigram tells us nothing about Ptolemy because it is almost surely not his writing. The key to the situation is that it occurs in only some of the manuscripts. Anyone who has worked with manuscript sources soon discovers that manuscript copyists work by rote. They have to: If they took time to think about the meaning of what they are copying, they would work more slowly and hence make less money. They copy everything before them, and leave out nothing except by inadvertence, and I shall give two examples.

Most Protestant Christians end the Lord's Prayer, the *Pater Noster*, with these words: "For Thine is the kingdom, and the power, and the

[11]Several writers say that Jonathan Swift rang the death knell of astrology when he wrote his parody *Prediction for the Year 1708* under the pseudonym Isaac Bickerstaff, Esq. I wish this were so. Unfortunately, as the current literature shows us, astrology is still a fervid faith with great numbers of the inhabitants of the United States, and probably of other countries as well.

[12]Some authorities say that Boethius translated the *Syntaxis* into ancient Latin in about the year 500. If so, this translation has been lost.

glory, for ever. Amen."[13] To me, it seems certain that these words were not part of the original manuscript of Matthew, nor of the original Christian tradition. One reason for thinking this is that they are absent from the corresponding place in Luke, as I have pointed out in a footnote. Another is that they do not appear in many versions of Matthew, and in particular they do not appear in the Vulgate, which is the translation of the Bible into Latin that was made by St. Jerome in about 385. What almost certainly happened was that some pious person added the extra words in the margin of a copy of Matthew and that a subsequent copier incorporated them into the main text. There are known cases with medieval documents where such a thing has happened.

The other example concerns the *Syntaxis* itself. One family of manuscript copies of the *Syntaxis* contains the records of six astronomical observations made by an astronomer named Heliodorus around 500, plus another observation that he made jointly with his brother Ammonius, and I have analyzed these observations elsewhere [*Newton*, 1976, Section V.9]. We know that these records could not have been written by Ptolemy even though they appear in some copies of the *Syntaxis*. The most likely explanation is that Heliodorus recorded his observations in his personal copy of the *Syntaxis*, and that this copy was the prototype of later copies, in which the copyists included everything before them, whether it could have been written by Ptolemy or not.

Going back to the quotation that opens some copies of the *Syntaxis*, it is highly unlikely that a copyist would omit the first sentence of what he is copying. Probably some admirer of the *Syntaxis* wrote this rhapsody at the head of his copy, and it was then included in copies made from this one. We should note that Manitius (see the entry for the *Syntaxis* in the list of references), in his German translation, puts this epigram on a separate page before the beginning of the main text, but he clearly does not include it as part of the *Syntaxis* itself. The English translation that is cited does not include it at all. Likewise it does not occur in either Halma's text or his translation, nor does it occur in Heiberg's text.

Thus we do not know from his own writing whether Ptolemy ever "tasted the intoxicating pleasure of a splendid theory" or not.[14]

[13]In the English King James Version, these words appear in Matthew 6:13, but they are absent from the corresponding point in Luke 11:4 where the prayer also appears.

[14]In his first footnote, Gingerich says that we do not know who the author of the epigram was, but he cites a paper by Franz Boll who "was inclined to attribute it to Ptolemy ..." I have not consulted the paper by Boll, but I see little possibility that Ptolemy was the author.

3. The Purposes of the Present Work

In the book that I designate as *CCP*, and in several earlier papers that are cited in it, I studied all the astronomical observations in the *Syntaxis*, whether they are attributed to other astronomers or whether Ptolemy claims to have made them himself. My conclusion [*CCP*, p. 344] was put in these words: "Thus we can say that all of Ptolemy's observations that he uses and that can be tested are fabricated." In addition, I concluded that many of the observations he attributes to others are in fact his own fabrications. It is clear that we cannot accept any statement made by Ptolemy unless it has independent confirmation.

As one might expect, several reviewers of *CCP* have disagreed with my conclusions, but I shall not attempt to cite all of them, because refutation of my critics is not the purpose of this work. In addition, *Gingerich* [1980] has published a paper that disagrees with my conclusions, and I gave a quotation from his paper near the end of the preceding section. Since Gingerich's paper makes many of the same points as the reviews that I have mentioned, I thought it worthwhile to write a commentary on his paper [*Newton*, 1980a]. In addition, *Swerdlow* [1979] has written a paper in which he disagrees with the conclusions of *CCP*.

I was stimulated to write the present work by a sentence from Gingerich's paper, which takes the form of the following question: "How can Ptolemy's parameters, which seem generally more accurate than his data base, be derived from observations that are simply fabricated?"

While refutation is not the purpose of this work, I think it is necessary to comment on this question immediately, lest I leave a wrong impression in the reader's mind. All the observations in the *Syntaxis*, whether they are attributed to Ptolemy or to others, agree with Ptolemy's parameters to a high level of precision that often amounts to exactness. This is what aroused my suspicions of Ptolemy in the first place. Therefore his parameters do not "seem generally more accurate than his data base ..." They are in fact just as accurate as his data base, no more and no less.

The answer to Gingerich's question is that positions calculated from Ptolemy's parameters, though not very accurate, still bear some resemblance to reality, and therefore they cannot be derived entirely from fabricated observations. It is therefore interesting to study the origin of Ptolemy's parameters, and doing so will be the main purpose of this work. This does not mean that I shall search through lesser known Greek and Babylonian sources in a quest for numerical values that Ptolemy later adopted. Rather, I shall study his computational and intellectual processes in order to find out the type of origin that he might have used.

In case this is confusing, let me give an example. The quantity that we call the maximum equation of the center[15] for the full moon is one of the parameters in Ptolemy's lunar theory. He adopts the value of 5° for this parameter, although no source known to us gives this exact value. I believe that Ptolemy adopted 5° because it is a round number that is close to results obtained by earlier astronomers. Thus I consider that the origin of this parameter is adequately explained, so far as the purposes of this work are concerned. However, it would still be an interesting historical problem to see if there are any earlier astronomers who also used exactly 5°.

A secondary purpose of this work is the following: The standard literature about the history of Greek astronomy contains a number of factual errors. In this work I shall correct the most important of these errors, particularly those that relate to our understanding of Ptolemy and his work.

In the current work I shall try as far as possible to avoid going over the same ground I covered in *CCP*. In spite of this intention, it will be necessary for me to repeat some of my earlier arguments, and I apologize to the reader for this repetition. Only in this way can I make it possible for the reader to study this work without constant reference to *CCP*.

4. Some Technical Matters

In writing dates in this work, I shall follow the astronomical rather than the historical practice. There are two points of distinction. One concerns the order of writing the elements of a date. I shall write the year first, followed by the English name of the month, and then the day of the month. As an example, I am writing this paragraph on 1980 August 2. In the main text I write the names of the months in full, but in tables, in order to have uniform columns, I use only the first three letters of the English name. Thus, for example, if I were writing the current date in a table, I would write it as 1980 Aug 2.

The other point of distinction concerns the years before the common era. I am writing this in the year 1980 of the common era. Let us count back to the year before the year 1 of the common era. In the historical style this year is known as the year 1 before the common era, but in astronomical style this is known as the year 0. The year before this is known as the year 2 before the common era in historical style and as the year −1 in astronomical style. Thus, for years before the year 1, the

[15]I shall define this confusing term in Section III.4. I say the full moon, because the value of the maximum equation of the center for the moon depends upon its phase.

number used for the year differs by 1 in the astronomical and historical styles.[16]

Ptolemy uses several ways to identify the date of an observation, but he always then translates it into the Egyptian calendar. The original record of an observation rarely uses an era. Instead, it may say, for example, that the observation was made in the 20th year of Hadrian. This means the 20th Egyptian year, not the 20th Julian year, of Hadrian's reign. When he comes to use an observation, however, Ptolemy always numbers the years from the first day of the Egyptian year in which the Babylonian king Nabonassar came to the throne. As Ptolemy uses it, this Egyptian "New Year" is −746 February 26, but we do not know if this date is accurate or not.[17]

I shall always translate Ptolemy's dates into the Julian calendar, with the years referred to the common era, but the reader should remember that Ptolemy did not do so.

By the closing period of Babylonian history, Babylonian mathematicians had developed the sexagesimal number system, with the use of a zero to indicate a missing position [*Neugebauer*, 1957, p. 16]. In writing a number in the sexagesimal notation, I shall use a semicolon to separate the fractional part of the number from the integral part, with commas separating the other positions. An example should make this clear:

$$11,37;20,0,48 = 11 \times 60^{1} + 37 \times 60^{0} + 20 \times 60^{-1} + 0 \times 60^{-2} + 48 \times 60^{-3}.$$

In the decimal system, we need 10 arbitrary symbols to represent zero and the first nine integers, and we express any other number by putting these symbols into the appropriate decimal positions. The reader may wonder whether the Babylonian schoolboy had to learn 60 arbitrary symbols and their associated values in order to write the values that appear in the various sexagesimal positions. By this I mean: Did he need, for example, separate symbols for 11, 37, and so on in the sexagesimal number written above?

The answer is no, he did not have to learn 60 arbitrary symbols. He expressed the values in the various positions, such as 11, 37, and so on, by a method that is like a simplified system of Roman numerals, with only two symbols. One looked somewhat like the Greek Γ, and I shall use

[16] It is unfortunate that some writers use the minus sign to denote years before the common era in historical style. For example, some writers use the symbol "−2" to denote the year 2 before the common era, which is properly denoted by the symbol "−1". Whenever the reader encounters a minus sign for the year in historical writing, he should check carefully to see what the writer means by it.

[17] That is, we do not know if the year is correct or not. This date is the first day of an Egyptian year.

Γ to represent the symbol in order to simplify the typesetting. The other looked somewhat like the symbol < that means "less than", and I shall also use this symbol for the sake of the typesetting. Γ represented 1 and < represented 10. Thus, to write the value 48 which appears in the last position above, a Babylonian wrote a cluster of four <'s followed by a cluster of eight Γ's.

Thus a Babylonian wrote the integers from 1 to 59 by a method that has an implicit decimal base, and he then used the sexagesimal system to represent all other numbers. Note that he used the symbol for 0 only in order to indicate a "missing" sexagesimal position, and that he did not use it in writing the value attached to a particular position. For example, to write the value 20 that appears in the first fractional position above, he wrote <<. No zero was needed.[18]

Ptolemy used the Babylonian system with two modifications. First, he used Greek numerals rather than Γ and <. The Greek number system was basically decimal, but without the zero, and I shall not take the space to describe how the Greeks managed without it. Thus Ptolemy needed only two symbols instead of twelve to write 48. Second, he wrote the entire integer part of a number in Greek numerals, which as I just said were decimal in structure. Thus, for the Babylonian number written above, Ptolemy would write 697;20,0,48. This system is almost the common system used today for writing angles.[19]

The fact that Ptolemy used a modified form of the Babylonian numerical system, and the fact that he "used" records of observations made in Babylon, does not necessarily mean that he could read the Babylonian cuneiform, for two reasons: (1) Many of the Babylonian observations that he used were fabricated, and all of them may have been [*CCP*, Table XIII.2]. (2) Alexandria was a center of astronomical activity centuries before the time of Ptolemy, and it is possible that earlier Alexandrian astronomers had translated many Babylonian records and left the translations in the library in Alexandria.[20]

18 This discussion of the Babylonian number system is primarily based upon the discussion by *Neugebauer* [1957, pp. 15ff].

19 The widespread use of digital computers (which ironically use binary arithmetic) may lead to the use of a fully decimal system. On the other hand it may not. My programmable pocket calculator has a "built-in" program for converting from decimal to sexagesimal form, and vice versa.

In modern astronomy, when we need to use a precision finer than the seconds position, we do so by expressing the number of seconds in decimal form. Thus we would write our example as 697;20,0.8 in modern notation. As *Neugebauer* [1957, p. 17] says, it has taken us 2000 years of development after the Babylonians to arrive at "a truly absurd numerical system" in which we write the largest and smallest components of a number in decimal form, with two sexagesimal positions in between.

20 There are reasons to believe that the library in Ptolemy's time was but a fraction of what it had been several centuries earlier. Still it is possible that the remaining part of the library contained the old astronomical data.

As a general but not inviolable rule, Ptolemy used his mixture of the Greek and Babylonian systems only in connection with calculations. In expressing the result of an observation, he generally used a different mixture, this time of the Greek and Egyptian number systems. In writing the integer part of a number he used the Greek system. In writing the fractional part of an observation he generally, but not always, used the Egyptian system of unit fractions. The only fractions that occur in Egyptian arithmetic, with one exception, are fractions whose numerator is 1; the exception is 2/3. A fraction was usually written by writing the symbol for the denominator, together with a special mark to indicate that the fraction rather than the integer was meant. I shall use a bar over a number for this purpose, and I shall use $\bar{3}'$ to indicate the fraction 2/3.

In the Egyptian system, every fraction had to be written as a sum of unit fractions, with it being understood that 2/3 is to be considered as a unit fraction in this context. Thus, for example, 3/5 had to be written as $\bar{2}\ \overline{10}$.[21]

With only a few exceptions, Ptolemy used only the following fractions in expressing the result of an observation: $\bar{6}$, $\bar{4}$, $\bar{3}$, $\bar{2}$, $\bar{3}'$, $\bar{2}\ \bar{4}$, and $\bar{2}\ \bar{3}$. In all cases, I believe, he wrote nothing at all when the fractional part was zero.[22] These fractions do not have a constant difference, and it is this fact which allows us to prove that Ptolemy simply appropriated his star catalogue from Hipparchus without acknowledgement. (See Section IX.7 of *CCP* or Section V.2 below.)

I cannot think of any instance in which we need to know how Ptolemy wrote a particular date, so far as the purposes of this work are concerned. However, there are times when it is important to know how Ptolemy wrote a particular number. Hence, while I shall write dates only in modern form, I shall usually write numbers in the form that Ptolemy did, except that I shall write fractions in modern form. It does not matter for present purposes that Ptolemy wrote the equivalent of $\bar{2}\ \bar{3}$ rather than 5/6. When I say that I shall write numbers in Ptolemy's form, I am referring to the structure used to write a number and not to the symbols used. Thus I shall use modern digits rather than the Greek ones.

We should note briefly the level of mathematics that was available to Ptolemy. Algebra was still centuries in the future, but both plane and solid geometry were well advanced. Ptolemy could solve both plane and spherical triangles in all the determinate cases that occur. However, the only trigonometric function that he knew is the chord function, which is

[21] The addition was understood; no sign was used for it. For further information about Egyptian mathematics, see *Neugebauer* [1957, Section 36].

[22] This applies only to writing the results of observations. In tables, which use the sexagesimal system for the fractions, the zero is explicitly written in when it applies, but it is omitted when Ptolemy uses the Egyptian fractions.

rarely used in modern mathematics, and this fact often makes Ptolemy's solutions for triangles rather cumbersome.

The chord function is illustrated in Figure I.1. In the figure,

$$\text{chord } \alpha = C/R.$$

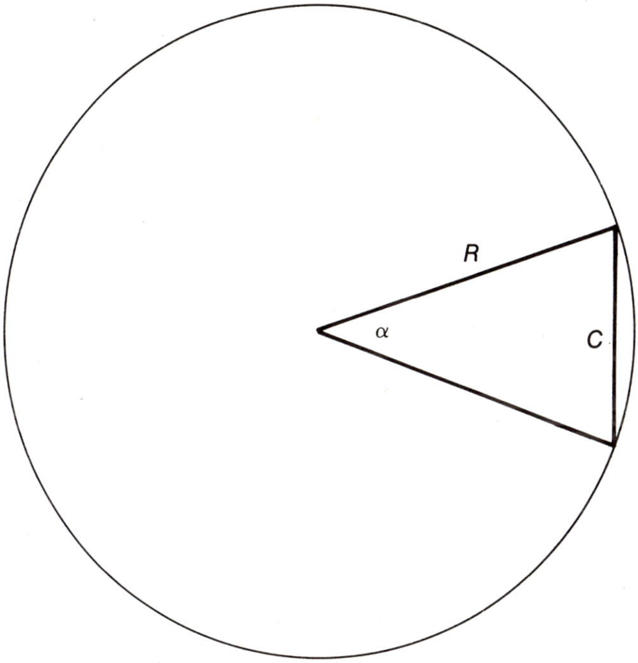

Fig. I.1—The chord function. The chord of the angle α is the ratio of C to R.

It is customary in modern mathematics to take the radius R of the circle to be 1, but Ptolemy usually takes it to be 60. If the reader will bisect the chord C, he will see that either half of it, divided by R, is the sine of $\alpha/2$. In other words,

$$\text{chord } \alpha = 2 \sin \tfrac{1}{2} \alpha. \tag{I.1}$$

We shall make frequent use of the concept of the mean longitude of a body. By the true or actual longitude of a body, I mean the angle measured in the plane of the sun's apparent orbit from the vernal equinox to the position of the body.[23] To a first approximation, every object in the solar system that concerns us moves uniformly in longitude. A point that moves uniformly in longitude, at the appropriate rate, and that always closely coincides with a body, is called the mean position of the body, or the mean body, and its longitude is called the mean longitude.

I shall use λ to denote the true longitude and L to denote the mean longitude of a body. The difference between λ and L is called the equation of the center of the body, and it will be denoted by e_c:

$$e_c = \lambda - L. \tag{I.2}$$

I can now complete the definition of the mean body: It is to be chosen so that the algebraic average of e_c is zero.

[23]If the body is not in this plane, its position has to be projected onto the plane. I shall not take the space to describe how this is done, and I imagine the reader's intuition will guide him sufficiently in this matter. I should also point out that true longitude in modern astronomy means something different from what I have said, but the current definition takes considerable space to develop, and I shall not give it.

CHAPTER II

THE SOLAR PARAMETERS

1. The Epicycle and Eccentric Models

It was a fundamental principle of Greek astronomy that all astronomical motions had to be compounded from uniform circular motions. This principle seems to have had about the same status with the Greeks as the principle of inertia has in modern physics. The simplest way to compound two uniform circular motions is furnished by the epicycle model, which is pictured in Figure II.1. We may safely assume that this model was known to Hipparchus around −140. Further, if a remark by Ptolemy is correct [*Syntaxis*, Chapter XII.1], the model was known to Apollonius of Perga about a century before Hipparchus, and possibly to even earlier astronomers. The basic principle of compounding uniform circular motions was not displaced from its central place in astronomy until the work of Kepler.

In Figure II.1 the point E represents the earth (the observer), and the line $E♈$ is some reference direction in the plane of the figure. At least since the time of Hipparchus, the line $E♈$ has been taken as the direction to the vernal equinox in most Greek, Islamic, and European astronomy. A point P moves at a uniform rate around a circle centered on E, and the angle L, increasing in the direction shown, is the mean longitude. The body of interest is represented by the point B, and it executes uniform circular motion around P. The angle γ is known as the anomaly. The angle $BE♈$, which is not drawn in, is the longitude of B.

The large circle centered on E is known as the deferent and the small circle centered on P is known as the epicycle. Since the angular motions given by the model do not depend upon the actual sizes of the deferent and epicycle, but only upon their ratio, the size of one circle may be chosen arbitrarily. In Greek astronomy it became standard to take the radius of the deferent to be 60, while the radius r of the epicycle was chosen to fit the circumstances.

In the most general case, the mean longitude L and the anomaly γ are independent but linear functions of the time t. That is:

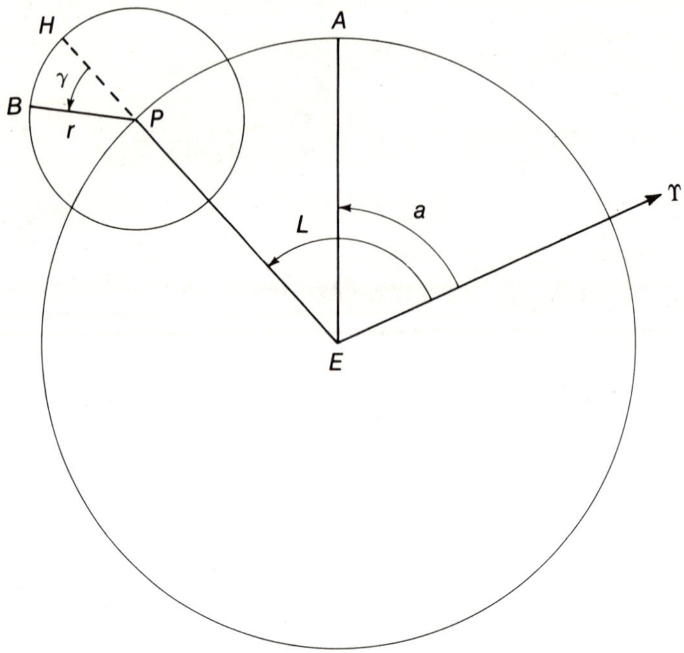

Fig. II.1—The simplest epicycle model. The line $E♈$ denotes the vernal equinox. A point P moves uniformly around the large circle while another point B moves uniformly around the small circle. B is the body whose motion is to be represented. The meanings of the other symbols are explained in the text. In some applications of the model, line BP is required to remain parallel to a reference direction, such as AE.

$$\begin{aligned} L &= L_0 + n(t - t_0) , \\ \gamma &= \gamma_0 + \gamma' (t - t_0) , \end{aligned} \qquad \text{(II.1)}$$

in which t_0 is a reference epoch that may be chosen arbitrarily. Equations II.1 contain four independent parameters, namely L_0, n, γ_0, and γ'. I shall not give names to these parameters for the moment.

The radius r of the epicycle is obviously another independent parameter, so the most general case has five free parameters. This case would arise if the model were applied to, for example, the motion of Venus. Since the orbit of the sun around the earth and the orbit of Venus around the sun are circular, to high accuracy, we can identify P with the sun and B with the planet, and have a rather accurate representation of the motion. The accuracy of the model for describing the motions of the other planets depends, in modern terms, upon the eccentricity of the planet's heliocentric orbit.

However, there is another use in which a restriction is placed upon the anomaly γ. In this use, the radius BP of the epicycle is constrained to be always parallel to some reference line such as AE, so that, for the sake of definiteness, the point B always lies directly above P in Figure II.1. If the reader will imagine that B is directly above P, he will see that the angles HPB and PEA are numerically equal, so that γ is always the negative of angle PEA, which equals $L - a$ in the figure. Since the Greeks had not yet developed the use of negative numbers, they changed the direction in which γ is measured in this case, so that γ would grow at the same rate as L.

When P is at the point A, it is clear that B has its maximum possible distance from E, so the point A is known as apogee, and the angle $AE♈$, which is denoted by a in the figure, is called the longitude of apogee. We clearly have the following relation among the angles L, γ, and a:

$$L = a + \gamma \,. \tag{II.2}$$

Further, a itself can be a linear function of time in Greek astronomy, so that

$$a = a_0 + a'(t - t_0) \,. \tag{II.3}$$

Aside from the radius r, we now have six parameters, namely L_0, n, γ_0, γ', a_0, and a'. However, Equations II.1 through II.3 tell us that there are two relations connecting them, so that only four are independent. Thus, somewhat surprisingly, imposing a constraint on the direction of the epicycle radius does not reduce the degrees of freedom. The freedom lost by the constraint is restored by the necessary introduction of the apogee direction AE.

The Greeks also had a way of producing an irregular motion by the use of only a single uniform circular motion. The secret of this apparent paradox lies in displacing the center of the circular motion from the position of the earth. The resulting model, called the eccentric model, is shown in Figure II.2. If we may trust the remark by Ptolemy that was mentioned a moment ago, the eccentric model was also known to Apollonius of Perga and perhaps to earlier astronomers.

In Figure II.2 the body in question is denoted by the point B, which moves uniformly around the circle centered on Z, but Z is displaced from the earth E by the eccentric distance e. The lines $Z♈$ and $E♈$ are parallel; both point to some reference point such as the vernal equinox. It is clear

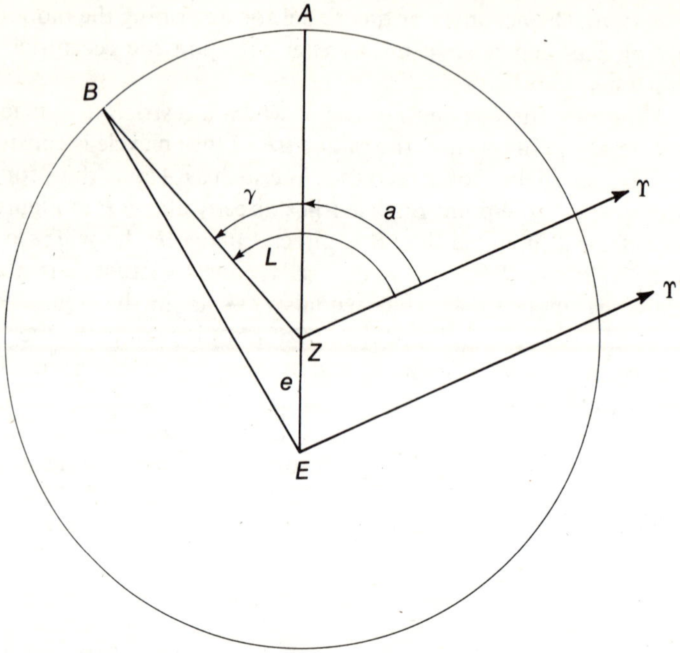

Fig. II.2—The simplest eccentric model. The body B whose motion is to be represented moves uniformly around the circle whose center is Z, but it is viewed from the eccentric point E, which is therefore the earth. The meanings of the other symbols are explained in the text. If the dimensions are suitably chosen, the eccentric and epicycle models give identical results.

that B is farthest from E when it is at point A, which is therefore the position of apogee in the eccentric model. Angle γ is again called the anomaly, while L is the mean longitude. It is clear from the figure that Equation II.2 still holds. The radius of the circle is again conventionally taken to be 60.

In fact, the epicycle and eccentric models are equivalent, provided that the epicycle radius r and the eccentric[1] e are equal. If the reader will imagine that point B in Figure II.1 is directly above point P, he will see

[1] I shall use the term "eccentric" to denote this distance in order to avoid confusing it with the modern term "eccentricity."

that the triangle *BPE* in Figure II.1 is congruent with the triangle *BZE* in Figure II.2 whenever the two values of L are equal. Thus, at all times, both models give the same value of the distance *BE* of the body from the earth, the same value of the anomaly γ, and the same longitude *BE*♈ of *B*. Hence, it is clear that we may make the same choice of free parameters in the eccentric model that we did in the epicycle model (that is, if we require *BP* to be parallel to *AE* in the latter case).

Ptolemy uses the eccentric model to describe the motion of the sun, and I shall follow his practice in this work.

2. The Number of Solar Parameters

As we saw in the preceding section, the eccentric model of Figure II.2 has five free parameters, provided that we take the radius R of the circle to be some arbitrary value such as 60. However, the number of parameters needed to describe the motion of the sun is greater than five in Greek astronomy.[2] To start with, we must locate the plane of Figure II.2, and this requires additional parameters.

Figure II.3 shows the parameters needed to locate an orbital plane, such as the plane of Figure II.2, in space. We start with the plane of the earth's equator, which passes through the center E of the earth. The reference line *E*♈ in this plane points toward the vernal equinox. The line *EN* is the earth's axis of rotation.

The orbital plane cuts the great circle labelled "orbit" from the celestial sphere, and the body B moves along its orbit in the sense indicated (all the astronomical bodies known to ancient astronomy moved eastward on the celestial sphere). We single out the point on the "orbit" circle where the projection of the body onto the celestial sphere passes through the equator in going from the south side to the north side. This is the point C in the figure. We use the term "ascending node" to denote the direction *EC*, and we use the symbol Ω to denote the angle ♈*EC*; this angle is taken to be positive when it is measured eastward, as it is in this figure.[3]

In dealing with the moon or the planets, we use the plane of the sun's orbit (the ecliptic) as the reference plane rather than the equator. In this case we use the symbol i, called the inclination, to denote the angle between the planes. When we are dealing with the sun, however, we use

[2]Because the Greeks did not have a theory of motion (dynamics) in the modern sense, many quantities that can be calculated in modern astronomical theory were free parameters to them.

[3]Sometimes we also use Ω to denote the direction *EC* itself, but the context always makes it clear which usage is meant.

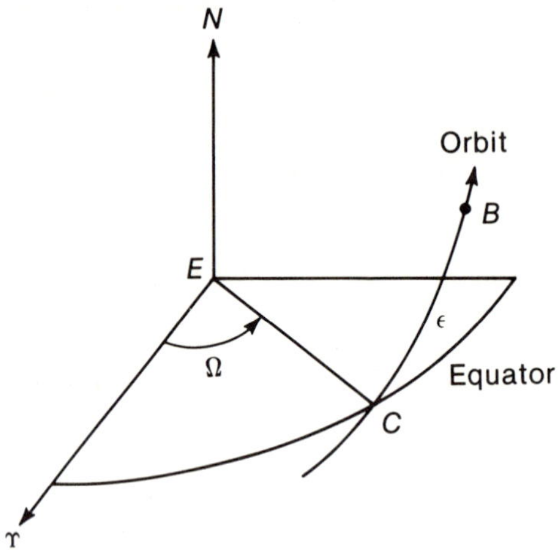

Fig. II.3—Orienting an orbital plane in space. The orbital plane of body B intersects the plane of the equator along a line EC, which is called the ascending node. The orientation can be specified by the angle Ω between $E♈$ and EC, and by the dihedral angle ϵ between the two planes.

the symbol ϵ, called the obliquity of the ecliptic, to denote the angle with the equator, as Figure II.3 shows. We usually shorten this name simply to the obliquity.

With regard to the nodal direction EC for the sun, we define the vernal equinox to be the direction of the sun's ascending node. Thus, by definition, C coincides with ♈ in Figure II.3, and the value of Ω for the sun is identically zero.

In summary of this point, only one free parameter is needed to locate the plane of the sun's orbit, but two parameters, the inclination and the node, are needed for any other body. The free parameter for the sun is the obliquity ϵ.

Up to now, we have been speaking of the radius of the circle in Figure II.2 as arbitrary, and we have assigned it the conventional value 60. However, various Greek astronomers made serious attempts to determine the actual value of the radius BZ. Hence, we must take the radius BZ, which I shall henceforth denote as R, as another free parameter. However, in all discussions in which we are interested only in angular relations, I shall continue to take the radius BZ equal to 60, and in particular I shall refer the eccentric e to this value.

We have also been speaking of the body as if it were a point, but even to the naked eye this is obviously not so for the sun and moon. In fact, the apparent sizes of the sun and moon are important in determining the circumstances of eclipses. Hence the apparent diameter d_S of the sun is another free parameter. The distance R and the apparent diameter d_S determine the physical size and volume of the sun.

Before closing this section, let us consider briefly the problem of measuring the obliquity ϵ. To do this, we measure the distance of the sun from the zenith at noon on the day of the summer solstice and also on the day of the winter solstice.[4] Then the obliquity is half the difference between these two values. The observer finds his latitude from these same data; it is the average (half the sum) of the two zenith distances. Since the observer's latitude comes from the measurements needed to find the obliquity, and since the observer must know his latitude in order to make sense of his observations, I shall take his latitude to be a free parameter, and I shall class it with the solar parameters for convenience.

We have now identified nine solar parameters that appear in Ptolemy's model of the sun, and I list them here for the convenience of the reader:

L_0,	the mean longitude at a reference epoch;
n,	the daily change in L, the mean longitude;
a_0,	the longitude of the apogee at the reference epoch;
a',	the rate of change in a;
e,	the eccentric distance (referred to 60);
ϵ,	the obliquity of the ecliptic;
β,	the latitude of the observer;
R,	the mean distance to the sun; and
d_S,	the apparent diameter of the sun.

Since we are interested in Ptolemy's treatment of the solar model, I shall take β to be the latitude of the place where he claims to have made his observations, which is Alexandria.

3. The Tropical Year

The list of parameters includes n, the daily change in the mean longitude L.[5] Since L is measured from the vernal equinox, the time in-

[4]In this discussion, I assume that both solstices come exactly at noon. Since this is not usually the case, some small corrections are needed to the procedure I am about to describe, but I think the general idea is clear.

[5]The time interval used to define n obviously does not have to be a day, but I shall follow Ptolemy and use the daily change.

terval required for L to change by 360° is the time required for the sun to proceed from one vernal equinox to the next, and this interval Y is clearly equal to $360°/n$. In other words, the time interval Y, defined by

$$Y = 360°/n \, , \tag{II.4}$$

is the length of the tropical year. Y and n are clearly equivalent parameters. n is more convenient to use in computations while Y is probably easier to think about.

When we come to consider the origins of Ptolemy's solar parameters, we shall see that the parameter he actually chooses is Y, and that he then derives the value of n, usually called the mean motion, by using Equation II.4.

Ptolemy claims [*Syntaxis*, Chapter III.1] that he measured the autumnal equinox in the year that we call 139, and that he measured both the vernal equinox and the summer solstice in 140.[6] He emphasizes the care taken in the measurement on two of these occasions. He then finds the length of the year in three ways. He compares his autumnal equinox in 139 with the corresponding measurement made by Hipparchus in −146, he compares his vernal equinox with the one that Hipparchus measured in −145, and he compares his summer solstice with one measured by Meton and Euctemon in −431. There is no doubt that Ptolemy fabricated the measurements that he claims to have made, and I have given the proof elsewhere [*CCP*, Section V.4].

Ironically, the summer solstice attributed to Meton and Euctemon is itself fabricated, although Ptolemy may not have known this. It was fabricated by some unknown person about the year −110, using the same solar theory (of Hipparchus) that Ptolemy used in his own fabrications. However, I do not think that this unknown fabricator had any intent to deceive. He simply needed the date of the solstice in −431 in order to put it on an inscription dedicated to Meton. See *CCP* [Section V.5]. Ptolemy may have fabricated the hour of the solstice himself, and the earlier person may have fabricated only the date.

It is interesting to digress for a moment and look at the accuracy with which Hipparchus and Ptolemy thought that Hipparchus had found the length of the year. We have three different values in the ancient

[6]With regard to the summer solstice in 140, Ptolemy uses a word that is usually translated as "compute" or "calculate", rather than a word that would be translated as "observe". However, he does not use this word to mean what we now mean by compute or calculate, and, as he uses it, it definitely implies observation. In order not to let this semantic question break into the main argument, I shall defer the question about the meaning of the word to Appendix A.

literature for the length that Hipparchus found. In discussing this accuracy, it will be convenient to compare Y, the year that Hipparchus found, with the length of the Julian year, which I shall denote by J. $J = 365\frac{1}{4}$ days.

Hipparchus found Y [*Syntaxis*, Chapter III.1] by starting with a measurement of the summer solstice made by Aristarchus in −279. He then calculated when the solstice would have occurred 145 years later, in the year −134, if the year had exactly the value J. By measurement, he found a discrepancy of "half a day and night together", that is, a discrepancy of 12 hours. Thus the discrepancy would be a full day in 290 years, leading to the relation

$$J - Y = 1/290 \text{ days} .$$

Ptolemy immediately rounds this to 1/300 days; this is the second value we find in the ancient literature. The third value is given by *Censorinus* [238, Chapter XVIII], who says that Hipparchus devised a calendar in which the average length of the year was $J - 1/304$.[7]

The three values attributed to Hipparchus are summarized in Table II.1. The total spread in the values of Y is about 0.000 16 days, or about 1 part in 2 million. I think there is little doubt that both Hipparchus and Ptolemy were willing to accept the differences between either 290 and 300 or 304 in the denominators. Thus they believed that the accuracy in Y is less than the equivalent of 7 decimal digits. Yet Ptolemy used Y to a precision of 6 sexagesimal positions, roughly equivalent to 11 decimal positions.

TABLE II.1

VALUES OF Y, THE LENGTH OF THE YEAR, ATTRIBUTED TO HIPPARCHUS

Source	$365\frac{1}{4} - Y$, *days*	Y, *days*
Data quoted by Ptolemy	1/290	365.246 55
Value used by Ptolemy	1/300	365.246 67
Value implied by Censorinus	1/304	365.246 71

[7] I think the number 304 comes from the fact that 19 years is an important interval in the lunar theory, and that 304 = 16 × 19.

In sum, Ptolemy essentially adopted the value of Y that had been found by Hipparchus. According to the data that Ptolemy quotes, the value that Hipparchus found was 365 + (1/4) − (1/290) days. Hipparchus is known to have changed the denominator to 304 for the purpose of calendrical computations. Ptolemy rounded the denominator to 300; we do not know whether Hipparchus ever used this specific alteration or not.

The value of n that corresponds to the adopted value of Y is

$$n = 0;59,8,17,13,12,31 \text{ degrees/day} . \qquad \text{(II.4a)}$$

4. The Sidereal Year

The tropical year means the time taken by the sun to pass from one equinox to another, while the sidereal year means the time taken between closest passages to the same star. Hipparchus discovered that these two years are not the same, and that the reason must be that the equinox is moving with respect to the fixed stars. That is, he discovered the precession of the equinoxes.

The tropical year, the sidereal year, and the rate of precession are interrelated, and only two of them are independent parameters. From Hipparchus on, most Greek, Islamic, and European astronomers have taken the tropical year and the rate of precession to be the more important parameters, with the sidereal year being relegated to a minor role.[8]

Ptolemy pays little attention to the sidereal year. In Chapter III.1 of the *Syntaxis* he remarks that the return of the sun with respect to the equinoxes and solstices is less than 365¼ days, while the return with respect to the fixed stars is more than 365¼ days. Hence, he says, Hipparchus concluded that the equinoctial points have a slow motion with respect to the stars. Ptolemy does not give the value of the sidereal year at this point, however, nor does he give any data from which it may be found. The only point where he gives data that lead to the sidereal year, so far as I can find, is in Chapter IV.2, where he is dealing with the moon. Here he says that Hipparchus found that the period of 126 007 days plus 1 hour is important in the lunar theory. This interval, according to Hipparchus, as quoted by Ptolemy, equals 4 267 synodic months,

[8] I believe that *Copernicus* [1543, Book III] is the most important later astronomer who preferred the sidereal year to the tropical year. He did so because he accepted Ptolemy's measurements of the equinoxes and solstices as genuine. Since Ptolemy's data relating to the tropical year did not agree with the data of other astronomers, whether ancient or medieval, Copernicus concluded that the tropical year was not a stable parameter, and he adopted the sidereal year instead.

4 573 anomalistic months (I shall define these terms in Section III.3), and 4 612 sidereal months less 7½ degrees, while at the same time it equals 345 sidereal years less the same 7½ degrees. However, Ptolemy does not calculate the sidereal year from these data, nor does he make any other use of the data that relate to the sun.

If the sun makes 345 revolutions, minus 7½ degrees, in 126 007 days plus 1 hour, the sidereal year is 365.259 859 days. Since Ptolemy does not take the sidereal year to be one of his parameters, this value does not concern us here. However, it will be important in Section V.3, where we take up the precession of the equinoxes.

5. The Apogee, the Eccentric, and the Anomaly

The data needed to find the sun's apogee a, its eccentric distance e, and the anomaly γ_0 at some epoch are the lengths of the seasons. *Geminus* [ca. −100] says that Hipparchus found that spring (the interval from vernal equinox to the summer solstice) has 94½ days, that summer has 92½ days, that autumn has 88⅛ days, and that winter has 90⅛ days. Since all the equinox and solstice measurements of Hipparchus that have been preserved are given only to a precision of a quarter of a day, it is unlikely that the values given for autumn and winter are observed ones. Probably Hipparchus added the measured lengths of spring and summer to get 187 days, and then subtracted this from 365¼ days to get 178¼ days for the combined length of autumn and winter. If this is right, he calculated the lengths of autumn and winter individually after he had found the parameters a, e, and γ_0, and Ptolemy does the same thing.

Ptolemy [*Syntaxis*, Chapter III.1] claims that he measured the autumnal equinox in 139 and the vernal equinox and summer solstice in 140, and that his measurements confirm these lengths of the seasons very closely. As we know, he fabricated his equinoxes by simply adding the appropriate multiples of Hipparchus's Y to Hipparchus's measured times for the equinoxes. He fabricated his solstice by adding the appropriate multiples of Y to the solstice attributed to Meton and Euctemon in −431. As I said above, we know this solstice was also fabricated by someone who used Hipparchus's solar theory, so the solstice fabricated by Ptolemy agreed accurately with Hipparchus's results, possibly to Ptolemy's great surprise.

Finding a, e, and γ_0 from the lengths of the seasons is a relatively rare problem that is easier to solve by geometry than by algebra. Ptolemy starts by finding a and e in Chapter III.4 of the *Syntaxis*. The figure that he draws to illustrate the solution, which the reader can find in *Pedersen*

[1974, p. 146], as well as in the texts and translations of the *Syntaxis*, is unnecessarily complicated and this makes it hard to follow his method. Figure II.4 is a simpler figure which contains only one line that is not used in finding *a* and *e*.

The circle in Figure II.4 is the eccentric circle of Figure II.2 with its center at *Z*, while the uniform motion of the sun around *Z* is viewed from the earth *E*. As in Figure II.2, I have put apogee at the top of the circle. Since spring was the longest season in ancient times, apogee lies somewhere between the vernal equinox and the summer solstice. In

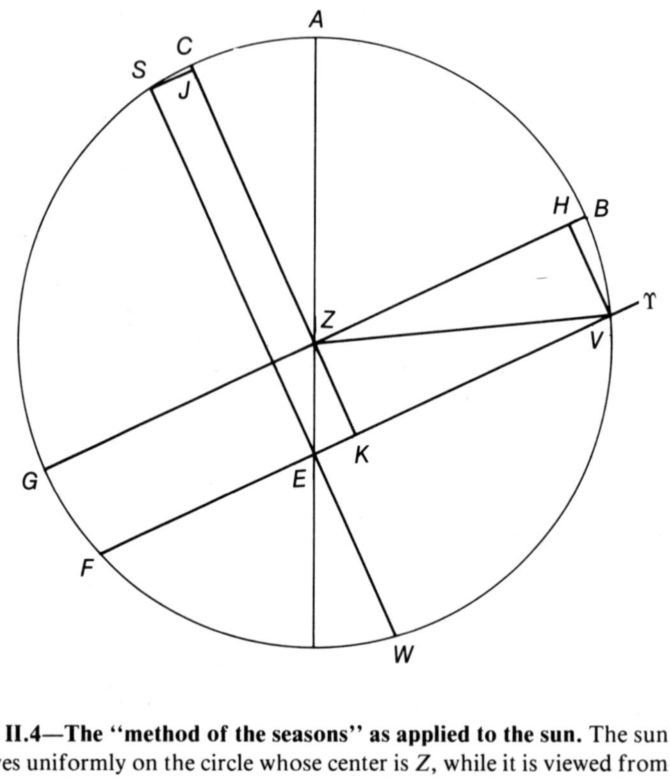

Fig. II.4—The "method of the seasons" as applied to the sun. The sun moves uniformly on the circle whose center is *Z*, while it is viewed from the eccentric point *E*. *E*♈ is the direction of the vernal equinox. At the vernal equinox, summer solstice, and autumnal equinox, the sun is at points *V*, *S*, and *F*, respectively. The lengths of the seasons give the lengths of the arcs *VS* and *SF*, and from them it is easy to solve for the triangle *ZEK*.

drawing Figure II.4 I have anticipated the result and put apogee 65;30 degrees from the direction *E*♈ to the vernal equinox.

The vernal equinox occurs when the sun, in its motion along the eccentric circle, comes to the point *V* that lies on the line *E*♈. Similarly, the summer solstice comes when the sun reaches the point *S* on the line *ES* that is perpendicular to *EV*♈. The autumnal equinox lies at the point *F* and the winter solstice lies at *W*. The line *EW* is not used in finding *a* and *e* in Ptolemy's method, which had already been used by Hipparchus and possibly by still earlier astronomers; only the lengths of spring and summer are used. Lines *BG* and *CK* in Figure II.4 are drawn through *Z*, the center of the circle, parallel to lines *FEV* and *SEW*, respectively.

During a time as short as the combined lengths of spring and summer, which is 187 days according to Hipparchus's measurements, we can neglect the change in the apogee and set a' equal to zero. This means that we can also set γ', the daily change in anomaly, equal to n, the daily change in the mean longitude.

The object in using Figure II.4 is to solve for the small right triangle *ZEK*, of which one angle is apogee *a*, and whose hypotenuse is the eccentric *e*.

The sun takes 94½ days to travel from *V* to *S*. If we use the value of *n* from Equation II.4a, the arc *VS* comes out to be 93;8,33 degrees, which Ptolemy rounds to 93;9. The sun takes 92½ days to travel from *S* to *F*, and the arc *SF* turns out to be 91;10,17 degrees, which Ptolemy oddly rounds to 91;11 rather than 91;10. The total arc from *V* to *F* is then 184;20.

Since *BG* is a diameter of the circle, the arc *BG* is 180°. Therefore the sum of the equal arcs *GF* and *BV* is 4;20, and arc *BV* equals 2;10 degrees.[9] In the triangle *HZV*, the angle at *Z* is 2;10 degrees and the hypotenuse *ZV* is 60, by convention. Hence the equal distances *HV* and *ZK* equal 2;16,6, which Ptolemy rounds to 2;16.

We have now found one leg of the right triangle *ZEK* and we find the other by a similar exercise. The arc from *V* to *S* is 93;9 degrees, but arc *BV* has just been found and arc *BC* is 90°. Thus arc *CS* is 0;59 degrees, and the leg *EK* of the triangle is 1;1,47, which Ptolemy rounds to 1;2. The eccentric *e* is the hypotenuse, which comes out to be 2;29,28 if we use 2;16 and 1;2 for the legs. Apogee *a*, which is the angle *ZEK*, comes out to be 65;29,33 degrees. Ptolemy rounds these values to 2;30, which equals 60/24, and to 65;30 degrees, respectively.[10]

[9]Ptolemy may have wanted the sum, namely 4;20 degrees, to be even so he could divide it by 2. He, or rather Hipparchus before him, may have chosen 91;11 rather than 91;10 for arc *SF* so that this would happen.

[10]Ptolemy actually uses angle *EZK*, which he takes to be 24;30 degrees, and thus he refers apogee to the summer solstice rather than to the equinox for some reason.

I shall call this method of finding a and e the "method of the seasons".

In order to find the anomaly γ at some epoch, Ptolemy, in his Chapter III.7, turns to the autumnal equinox which he claims to have measured in the year 132. To find γ at the time of this equinox, we turn to Figure II.5, which again is much simpler than the figure Ptolemy draws. The sun is at point F. In the triangle FZE we know the sides FZ and ZE; the latter is the eccentric e. We also know the angle β, because it is $180° - a = 114;30$ degrees. The angle e_c is the equation of the center.[11] When we apply the law of sines to the triangle, we get $\sin e_c = (e/60) \sin \beta$. This gives $e_c = 2;10,22$, which Ptolemy rounds to 2;10 degrees.

γ is clearly the sum of e_c and β, so it equals 116;40 degrees, and the mean longitude L equals 182;10 degrees.

Now let us ask how Ptolemy chose the parameters a, e, and L_0. Since he shows himself interested in maintaining continuity between his results and those of Hipparchus, it is natural that he would choose to keep Hipparchus's e, which was 2;30 for a mean orbital radius of 60. The interesting point concerns his choice for a, which interacts with the value of the anomaly at the epoch of his autumnal equinox of 132. Hipparchus had also found a to be 65;30 degrees.

Here we must anticipate Ptolemy's work with the planets. As we shall see in Chapters VII and VIII, Ptolemy has to choose a value of a' for each planet, and he chooses to make a' exactly equal to 1° per century for each of them. We shall see the reason for this choice in Section VII.2. For the sun, however, Ptolemy chooses to make a' exactly zero. That is, he chooses to keep the sun's apogee always at the same angle from the vernal equinox, so that the sun's orbit precesses slowly with respect to the stars.

Since a' is zero for the sun, we do not need to speak of the value of apogee a_0 at some chosen epoch, and I shall simply speak of its value a as I have been doing (thus anticipating the value of a' that Ptolemy chose).

In the rest of this section, I shall give two possible reasons why Ptolemy chose to make a' equal to zero. In order to understand the first reason, we must consider how Ptolemy would fabricate the time of an equinox or solstice for given values of the parameters, if the position of apogee is not constant.

To find the time of the autumnal equinox, for example, he would start with Figure II.5. Once he has chosen a, he would subtract it from 180° to get the angle β in the figure. Then he would get the anomaly γ in the same way that he did for the autumnal equinox of 132. Adding this to

[11] Strictly speaking, the equation of the center is negative in the situation shown, so the angle marked e_c is actually the negative of the equation.

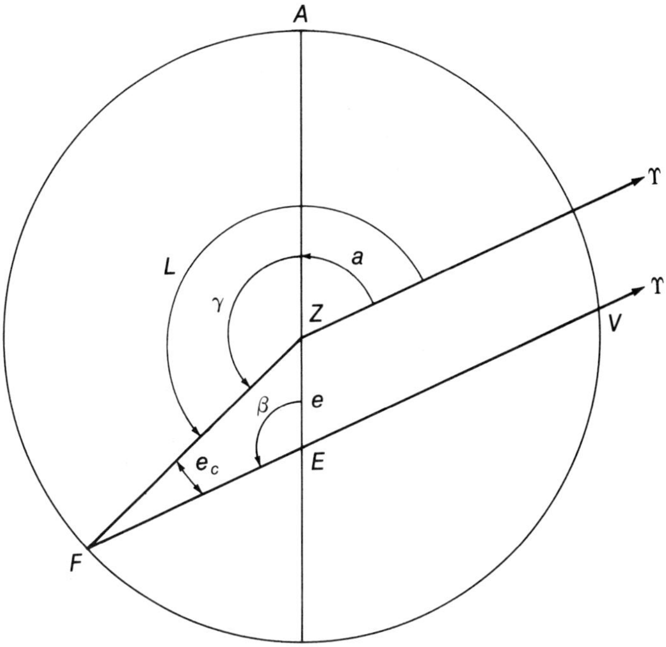

Fig. II.5—Finding the anomaly and mean longitude of the sun at an autumnal equinox. The angle β and the distance e are known from Figure II.4, and the distance FZ is known by convention. Thus the triangle FZE can be solved, and from it we can get the angles γ and L.

a would give him the mean longitude L, and from this he could calculate the time. To find the time of a summer solstice, he would do the same thing, except that β would be $90° - a$; and, for a vernal equinox, β would be $0° - a$, or what is the same thing, $360° - a$.

After he had found the three times, he would then have to turn around and solve for a, e, and γ by the process we have already been through.

However, if he kept a always in the same place, his operations were much easier. To fabricate the times of the solstices and equinoxes, he had only to add the appropriate multiple of the length of the year to the equinox and solstice times that Hipparchus had measured. Since this gave the same lengths of the seasons that Hipparchus had used, Ptolemy did not have to solve for a, e, and γ; he could simply copy the solution already given by Hipparchus.[12]

There is at least one other possible explanation of why Ptolemy chose to make $a' = 0$. The equinoxes are intimately connected with the solar motion but they play little role in the planetary motions except to serve as convenient reference points. Thus Ptolemy may have thought that it made a more plausible cosmos if the solar orbit stayed fixed with respect to the equinoxes while the planetary orbits stayed fixed with respect to the stars.

We have no way to know which explanation is correct, and, since they in fact reinforce each other, both may have played a part in Ptolemy's thinking. Thus we do not have to make a choice between them, since either one by itself is adequate and since both lead to the same result. Nonetheless, I have a personal preference for the first one, because it seems to me to be consistent with Ptolemy's methods of working. We shall see other instances in which it is almost certain that Ptolemy made a particular choice of parameters for the sake of "economy of effort".

6. The Mean Longitude and the Anomaly at the Epoch

For his fundamental epoch, Ptolemy chooses noon, apparent time at Alexandria, on −746 February 26.[13] Ptolemy takes this to be the first day of the Egyptian year in which Nabonassar became king of Babylon. As I have already remarked in Section I.4, we do not know whether this date is historically correct or not, but this question, though important for history, does not matter for astronomy. All we need for astronomy is a specific date in a calendar based upon known principles. Ptolemy's fundamental epoch is usually called the era of Nabonassar in modern writing. Contrary to many statements in the literature, there is no evidence that it was ever used for any purpose except making astronomical calculations.

To find the mean longitude L_0 of the sun at the era of Nabonassar, Ptolemy uses the value of L that we found in the preceding section by using the autumnal equinox of 132. According to Ptolemy, this came at 2 hours after noon on September 25, and he emphasizes the care with which he made the measurement. Actually we know [*CCP*, pp. 90ff]

[12] The reader may well ask: Why did Ptolemy not take $a' = 0$ for the planets also? Would he not have saved himself some labor by doing so? The answer is that he would not, but I cannot show this until we come to the treatment of the planets in Chapters VII and VIII.

[13] In an earlier work [*Newton*, 1976, p. 43] I accidentally wrote the Julian day number of this date as 1 448 368. It should have been 1 448 638.

that he fabricated this measurement by adding 278 Y (Y is still Hipparchus's length of the year) to the measurement of the autumnal equinox that Hipparchus made in -146, and then rounding the result to the nearest hour. Thus the value of L_0 that Ptolemy adopts is basically Hipparchus's, although he may have used a different era from Hipparchus. The value of L_0 that he finds is 330;45 degrees.

The value of γ that we found in the preceding section applies at the epoch of the autumnal equinox in 132. In his tables and computations, Ptolemy uses a value referred, like L_0, to the era of Nabonassar. Since $L_0 = 330;45$ degrees, while a is constant at 65;30 degrees in Ptolemy's theory, it follows that γ_0, referred to the era of Nabonassar, is 265;15 degrees.

7. The Obliquity and the Latitude of Alexandria

In Chapter I.12 of the *Syntaxis* Ptolemy describes two meridian instruments that he claims to have used in measuring the zenith distance of the sun. One is a full circle while the other is a quadrant, but both give equivalent information. He says in Chapter III.1 that he used these instruments to measure equinox times. He does not say how he measured the solstice, except to say that he measured it "very carefully", but his claimed solstice measurement occurs in the same context and the implication is that he measured it also with a meridian circle or quadrant.

In Chapter I.12 Ptolemy claims that he used these instruments to measure the obliquity of the ecliptic, by the basic method I mentioned in Section II.2. He says that he measured the angle between the tropics several times, and that he found it always to lie between 47;40 and 47;45 degrees. He goes on to say that this is almost the same value as that of Eratosthenes, which is also the value that Hipparchus used, namely 11/83 of a full circle. Now the average of the limits that Ptolemy claims to have measured is 47;42,30 degrees while 11/83 of a circle is 47;42,39 degrees.

On the basis of these results, Ptolemy takes the obliquity ϵ to be 23;51,20 degrees. The correct value in his time [*CCP*, p. 97] was about 23;40,44 degrees, more than 10′ less than the value Ptolemy found. It is not conceivable that he always measured a value that agreed accurately with Eratosthenes' value but that was in error by a large amount. Thus his data must be fabricated.[14]

Incidentally, Hipparchus's solar observations suggest that he never measured the obliquity himself, and that he simply accepted

[14]Since the obliquity is found from the difference of two measured zenith distances, it is free from most instrumental bias. It is still affected by refraction, but this contributes an error of only about 40″.

Eratosthenes' value. The interval that he measured between a vernal equinox and the succeeding autumnal one is too long by about 0.6 days [*CCP*, p. 82]. Aside from small accidental errors of observation, the errors in vernal and autumnal equinoxes tend to be equal and opposite. Thus Hipparchus's error is about 0.3 days in each equinox, and the event that he called the equinox came when the sun was south of the equator. In 0.3 days, the declination of the sun changes by about 7′, so Hipparchus's equinoxes came when the sun was about 7′ south of the equator.

This suggests [*CCP*, pp. 99ff] that Hipparchus first measured the zenith distance of the sun (or its elevation angle) at the summer solstice. Then, if this idea is correct, he took the equator to be south of the summer solstice by 23° 51′ 20″, which is Eratosthenes' value, and which is too large by about 8′. Further, according to this idea, he never measured the winter solstice at all, and this is consistent with the way that he stated the lengths of the seasons involving it (Section II.5 above). However, this idea apparently conflicts with the conclusions of *Rawlins* [in press] drawn from a study of Hipparchus's star catalogue. In fact, we must study the star catalogue before we can decide why Ptolemy chose 23° 51′ 20″ rather than some other value for the obliquity. I shall do this study in Section V.4.

Returning to Ptolemy's solar observations, the latitude of the observer is derived from the same data that yield the obliquity, as I remarked in Section II.2 above. However, Ptolemy does not derive his latitude in Chapter I.12, where he finds the obliquity. He does not mention his latitude until Chapter V.12, where he uses it in the process of finding the moon's parallax. There he says that he made his observations "in the parallel of Alexandria" and that the latitude β of that parallel was 30;58.

If Ptolemy really made his observations "in the parallel of Alexandria", his latitude was [*Times Atlas*, 1955] about 31;13 degrees. If he really made them at Canopus, his latitude was about 31;19 degrees. Thus his error in his latitude was at least 15′. It is not credible that he could have made as many observations of the sun as he claims to have made without finding his latitute more accurately than this, and we conclude that his claimed measurement of the latitude was fabricated.

I have already [*Newton*, 1980] discussed the origins of the obliquity ϵ and the latitude β in some detail, and I have shown that the value of β, as well as the value of ϵ, may go back to Eratosthenes or to earlier astronomers, even though Ptolemy does not say anything about an earlier value of β. *Rawlins* [in preparation] has given us the clue that we need to understand these origins. He starts from the fact that the latitudes given in ancient geographies are often in error by a degree or more, even though it should have been easy to measure them with an accuracy of about 5′. However, Greek astronomers and geographers did not originally use latitude as the north-south coordinate of a point.

Before they used the latitude, they sometimes used the length of the longest day, and they sometimes used the noontime length of the shadow cast by a vertical rod of a standard length (called a gnomon) at the equinoxes. The latter value is, in modern terms, proportional to the tangent of the latitude. The values of these coordinates were naturally rounded to convenient values. Later, when latitude came into common use, the latitude was often calculated from these earlier rounded coordinates instead of being measured afresh.

If we take 31;13 degrees for the latitude of Alexandria, its tangent is 0.606 019. It would have seemed quite reasonable to an ancient geographer (or astrologer) to round this to 0.6 exactly, and $\tan^{-1} 0.6 = 30;57,30$ degrees.[15] I feel sure that this is the ultimate reason why Ptolemy took β to be 30;58 degrees, although he probably was not the first person to use this value.

Now let us ask: If the declination of the sun is δ, and if an observer is at latitude ϕ, what is the length of the day? In asking this question, we neglect the finite radius of the sun and we neglect refraction. That is, we are asking a purely trigonometrical question. We can answer the question by using the transformations given in Section 2A of the *Explanatory Supplement* [1961]. The answer is:

$$\text{length of day} = 2 \cos^{-1} (-\tan \delta \tan \phi) \ . \qquad \text{(II.5)}$$

Next we modify the question to read: What is the length of the longest day for a point on the Tropic of Cancer? On the longest day, the declination of the sun equals the obliquity ϵ, and on the Tropic of Cancer the latitude ϕ also equals ϵ. Hence the longest day on the tropic is given by:

$$\text{longest day} = 2 \cos^{-1} (-\tan^2 \epsilon) \ . \qquad \text{(II.5a)}$$

In Chapter II.6 of the *Syntaxis* Ptolemy gives a table of the latitudes at which the longest day takes on various specified values, and he uses $\epsilon = 23;51,20$ degrees in the calculations. For a latitude of 23;51 degrees[16] he finds that the longest day is exactly 13½ hours. Actually the longest day, for the values Ptolemy uses, is 13;30,11 hours rather than 13;30. If we take the longest day on the Tropic to be 13;30 hours exactly and solve for ϵ, we find 23;49,50 degrees rather than 23;51,20.

I feel there is little doubt that the length of the longest day on the tropics was believed to be 13½ hours, and this is exactly the value that

[15] I do not mean to imply that the tangent function was used by the ancient Greeks. I am merely using the tangent for brevity to express a value that they could express only in more complicated terms.

[16] Ptolemy rounds all the latitudes to the nearest minute in this table. I believe that he means for this latitude to be the same as the obliquity.

Ptolemy gives. However, this does not lead to exactly the obliquity that Ptolemy uses, and that Eratosthenes had used before him. But at least it explains why Eratosthenes' famous value of ϵ was considerably too large. Taking 23;51,20 degrees for the obliquity rather than 23;49,50 requires that we postulate a slight error in computation, in addition to taking the longest day to be exactly 13½ hours.

This would sound like an improbable origin for Ptolemy's value of the obliquity except for the fact that we see the error in computation actually displayed in the *Syntaxis*. The error does not arise from errors in Ptolemy's trigonometric table. I have recast the problem entirely in terms of the chord function, which is the only trigonometric function that Ptolemy knew, and I have solved the problem using Ptolemy's own table of the chord function. I got exactly the values just given, to the accuracy with which they are stated.

I believe that a possible explanation of the situation is the following: It was accepted by the ancient Greek astronomers that the length of the longest day on the tropics was 13½ hours. Then someone, perhaps Eratosthenes, decided to calculate the obliquity from this value. Either he made a simple mistake in calculation, or he had trigonometric tables that were not as accurate as those of Ptolemy. As a result he found ϵ = 23;51,20 degrees rather than 23;49,50.

I doubt that Ptolemy actually calculated the latitude for which the longest day is exactly 13½ hours. I think he simply accepted a long-taught value without checking it. Further, I think the entire table of latitude as a function of the longest day was already several centuries old in his own time, and that he also accepted it without examination. I shall give the reasons for this conclusion in Appendix D.

Moesgaard [1980] has a different explanation for the origin of the value 23;51,20 degrees.[17] He concludes that Babylonian astronomers inferred the obliquity from lunar eclipses, that their method inherently gave a value that is too large, and that the Greek astronomers adopted their value instead of measuring it themselves. He notes [p. 83] that this argument by itself does not produce the exact value in question, but he finds a numerological basis for it: It comes from requiring the meridian shadows of a gnomon at the soltices and the equinoxes to be in the ratios $2^3:3^3:4^3$. This in turn requires the latitude ϕ and the obliquity ϵ to satisfy the following equations:

$$\tan(\phi - \epsilon) : \tan\phi : \tan(\phi + \epsilon) = 2^3 : 3^3 : 4^3 \; .$$

It seems mildly unlikely that these equations can be solved in closed form, not to mention in rational form, but they can be, and the solutions

[17] I thank Professor Moesgaard for sending me an advance copy of the page proof of his paper.

are $\tan^2 \epsilon = 115/588$ and $\tan^2 \phi = 243/460$. The corresponding angular values are $\epsilon = 23;51,25$ and $\phi = 36;0,37$. *Rawlins* [in preparation] has shown that a common way to express an angle in Greek mathematics was to take its ratio to a full circle or a half circle, and then to expand the result as a continued fraction. If we let $\alpha = \epsilon/360°$, the continued fraction for α is

$$\alpha = \cfrac{1}{15 + \cfrac{1}{11 + \cfrac{1}{8 + \cdots}}}.$$

The first approximation gives $\alpha = 1/15$, leading to $\epsilon = 24;0,0$ degrees. The second approximation gives $\alpha = 11/166$, leading to $\epsilon = 23;51,20$ degrees. This is the famous value of α attributed to Eratosthenes. Further, Ptolemy [*Syntaxis*, Chapter I.12] specifically says that Eratosthenes expressed his value as 11/83 of a half circle, which is identical with the value of α that we get from the second approximation.[18]

Certainly some ancient Greek philosophers (including some whom we may call astronomers) believed that the universe was controlled by number, and thus Moesgåard's hypothesis is within the general framework of ancient thought, even though there is no specific example of such a complicated numerology that is known to me. My main objection to the hypothesis is the following: If Eratosthenes found the ratio 11/83 in the way that is postulated, this means that he could find the inverse of a trigonometric function with an accuracy of a few seconds of arc. Yet he is only a few decades later than Aristarchus of Samos, who could say of cos 87° only that it lies between 1/18 and 1/20 (next section).

If Eratosthenes could indeed find trigonometric functions with the accuracy that Moesgaard's hypothesis requires, why should he make the error of taking the solution of Equation II.5a to be 23° 51′ 20″ when the length of the longest day is taken to be exactly 13½ hours? Any answer to this problem is necessarily tentative, but the following seems to be the most probable on the basis of current knowledge: The accuracy of trigonometric tables increased greatly between Aristarchus and Eratosthenes, to the point that the latter could get 23° 51′ 20″ rather than 23° 49′ 50″ for the solution of Equation II.5a. However, he would

[18]The third approximation is $\alpha = 89/1343$, which leads to $\epsilon = 23;51,25$ degrees. This is the same as the exact solution, to the nearest second of arc.

not have been able to get the highly accurate solution required by Moesgaard's hypothesis.

In summary, it does not seem likely that the standard Greek value for the obliquity, which Ptolemy claims to have verified by observation, was the result of careful observation at all. It was probably "numerological" in origin, and we have two suggested origins for it. One suggestion requires an error in calculating the length of the longest day on the tropics, but we see this error explicitly displayed in the *Syntaxis*, and thus this requirement is not an argument against this suggestion. The other suggestion leads accurately to the value in question, but it requires either an early knowledge of accurate trigonometric functions or an error in tradition. I shall take up the choice between the two suggestions in Appendix D, where I conclude that the first one is more likely.

Incidentally, the fractions that result from the technique of continued fractions are quite sensitive to the quantity being represented. We saw above that the value of ϵ that leads to 13½ days for the longest day on the tropics is actually 23;49,50. If we expand the ratio of this to 360° as a continued fraction, the result is

$$\cfrac{1}{15 + \cfrac{1}{9 + \cfrac{1}{3 + ..}}}$$

The second approximation that results from this is 9/136, which leads to ϵ = 23;49,25. *A fortiori*, the use of a value that is close to the correct ϵ would yield a fraction whose members differ even more from the classic 11/166.

8. The Distance and Apparent Diameter of the Sun

With the methods he uses, Ptolemy has to find the distances of the sun and moon simultaneously, and the same thing applies to their apparent diameters. Hence we cannot complete the discussion of these values until we come to Chapter IV. However, in order to find the solar and lunar values simultaneously, Ptolemy first establishes some ratios which we can study here. In fact, these ratios are fundamental parameters in Ptolemy's procedures, and the solar distance and diameter are derived from them.

First let us consider the apparent diameter of the sun. Ptolemy takes up this topic in Chapter V.14 of the *Syntaxis*. He says that he has built an instrument for measuring the apparent diameters of the sun and moon that is like one Hipparchus had described, and that he (Ptolemy) has made observations with it.[19] This instrument consisted of a rigid base which had a perpendicular plate with a pinhole in it at one end. Another perpendicular plate, which could slide along the base, carried a square. This slide was moved along until the circle of the sun or moon appeared to be inscribed in the square. I presume that the distance between the pinhole and the square was read on a scale and the apparent diameter determined from this distance and the size of the square. This instrument is called a *dioptra*.

Ptolemy says that the *dioptra* can be used to make accurate comparisons of the solar and lunar diameters, but that it cannot furnish accurate values of either. What he says on this point does not make sense to me in any of the translations (French, German, and English) that I have read, but I do not think this is the fault of the translators.[20] Rather, I think, it is not possible to uphold this position and make sense. Of the various methods that Ptolemy describes for finding the diameters, use of the *dioptra* should be the best rather than the worst. Certainly it is better than the method that Ptolemy adopts, as we shall see in Section IV.2.

Using this instrument to compare the diameters of the sun and moon, which Ptolemy says it can do quite well, he finds first that the sun's diameter shows no appreciable variation, although it actually varies from about 31′ 31″ to about 32′ 35″. He also says, still on the basis of his observations, that the (constant) diameter of the sun is the same as the minimum diameter of the moon. He points out that this disagrees with the results of earlier astronomers, who found that the solar diameter is equal to the average, not to the minimum, lunar diameter.

One consequence of Ptolemy's findings is immediately contrary to observation. It denies the possibility of an annular eclipse, yet slightly more than half of all central eclipses are annular. Furthermore, the existence of annular eclipses had long been known to Greek astronomers, if a passage from Simplicius is correct.

Simplicius was a philosopher from the eastern Roman empire who flourished in the early 6th century, and some of his writing is quoted by

[19]On page 171 of the English translation of the *Syntaxis* that is cited in the references, the translator Taliaferro gives a description of this instrument taken from a commentary on the *Syntaxis* written about the year 300.

[20]Further, a Greek friend of mine (S. M. Krimigis) has read the original at my request and he says that it does not make sense either.

Heath [1913, p. 222] ; I have not consulted an edition or translation of the original. In the passage in question, Simplicius quotes from an astronomer named Sosigenes, presumably the Sosigenes who advised Julius Caesar on the occasion of establishing the Julian calendar. If so, the quotation is about two centuries earlier than Ptolemy. It runs: "... when the centre of the sun, the centre of the moon, and our eye happen to be in a straight line, what is seen is not always alike; but at one time the cone which comprehends the moon and has its vertex at our eye comprehends the sun itself at the same time, and the sun even remains invisible to us for a certain time, while again at another time this is so far from being the case that a certain rim of a certain breadth on the outside edge is left visible all round it at the middle of the duration of the eclipse." If this is a valid quotation from Sosigenes, annular eclipses were known to Greek astronomers at least two centuries before Ptolemy, and he was indeed temerarious (but not necessarily ill-informed) in denying their existence.

The earlier astronomers were more accurate than Ptolemy about the apparent sizes of the sun and moon. The average lunar diameter is about 31′ 6″ while the average solar diameter is about 32′ 2″. The minimum lunar diameter is about 29′ 24″, much smaller than the average solar diameter. I believe this is the only instance in the *Syntaxis* in which Ptolemy deliberately chose to differ significantly from accepted results, and we naturally ask why.

I also believe this is the only choice that Ptolemy makes in the *Syntaxis* for which I have no satisfactory explanation. I present a speculative explanation on page 201 of *CCP*, based on writings of Ptolemy other than the *Syntaxis*. Even if the discussion I give there is correct, it merely says that Ptolemy had to make a choice between two possibilities at one point in his work, but it does not tell us why he made the choice that he did. Perhaps his choice was arbitrary. I shall return to this problem in Section IV.6; all I can say here is that the matter needs more research.

Now let us turn to the ratio of the solar and lunar distances. So far as we know, Aristarchus of Samos [*Aristarchus*, ca. −280] was the first person to estimate this ratio. He did so by observing when the moon seems to be exactly at a quarter, that is, when exactly half of the moon's disk seems to be illuminated. At this time, as a little thought shows us, the earth-moon direction and the moon-sun direction must be perpendicular. Let the angle between the sun and moon as seen from the earth be denoted by D; D is called the lunar elongation. Then $\cos D$ is the ratio of the lunar distance to the solar distance. Aristarchus took D to be 87°, which makes the ratio of the distances be 1/19.1, while the correct ratio is close to 1/400. Aristarchus was apparently too early to have accurate values for the trigonometric functions, because he says only that the ratio is between 1/18 and 1/20.[21]

Ptolemy finds the solar and lunar distances by a method that does not apparently use this ratio. However, the data that he uses to find the distances are clearly fabricated, and the distances that he finds have a ratio that is close to 19. Thus I think that he fabricated his data in a way that would preserve Aristarchus's already classic value.

9. The Accuracy of Greek Measurements of Solstices

There is an interesting point about the accuracy with which a Greek astronomer could measure the time of a solstice. Properly, this topic belongs earlier, in Section II.3, but it is a lengthy topic and I have put it here in order not to interfere with the development of other arguments.

In Section V.2 of *CCP* I studied the accuracy of Greek measurements of equinoxes and solstices. We expect an equinox to be more accurate because the declination of the sun is changing most rapidly at that time. However, the result was that both measurements had about the same standard deviation, namely about 7 hours. The explanation of this situation is that an equinox can be measured with more precision than a solstice, but the equinoxes are subject to various sources of bias that do not affect the solstices. Thus the overall accuracy turns out to be about the same for both.

The solstice time that Ptolemy claims to have measured is late by 36 hours, about 5 standard deviations, but it agrees exactly, at the level of rounding used (which is to the nearest hour), with Hipparchus's theory of the sun. Similar statements are true of the equinoxes. Thus, I see no way to escape the conclusion that Ptolemy fabricated these data.

Swerdlow [1979, p. 527] does not agree with this conclusion, and in particular he does not feel that an error of 36 hours in measuring the time of a solstice is unreasonable. He expresses the situation in this way: "At the time of solstice, the meridian altitude of the sun changes by less than fourteen seconds of arc per day, and measuring this quantity, let alone any fraction of it, was obviously ridiculous."

In writing this, Swerdlow has apparently overlooked the study of solstice accuracy in *CCP*, which was based upon actual Greek observations, as well as the study of *Hartner* [1977] on the same subject. He has also apparently overlooked the fact that the solstice can be defined in two

[21] Heath, on pp. 333ff of the cited edition of *Aristarchus*, says that he knew certain limits for the trigonometric functions, one of which is that sin 90°/m lies between 3/2m and 5/3m, provided that m is greater than 4. Heath is drawing upon a paper by P. Tannery published in 1883 that I have not consulted.

different ways: (a) The solstice is the time when the meridian altitude of the sun is an extremum, or (b) the solstice is the time about which the changes in meridian altitude are symmetric. Swerdlow's objection applies if we use the first definition, but not if we use the second.

In this section, I want to estimate the accuracy of finding a solstice if we use the second definition. We must start by finding the precision with which a Greek astronomer could measure the meridian elevation of the sun. Ptolemy [*Syntaxis*, Chapter III.1] quotes 14 measurements of the vernal equinox made by Hipparchus over a time span of 18 years, and we have good reason to believe that they are genuine. The times are rounded to the nearest multiple of 6 hours, and all are consistent with a year of 365¼ days. This implies that Hipparchus's precision (but not his accuracy) was about 2′ of arc.

However, Hipparchus was probably more precise than the average Greek astronomer. To be conservative, I shall use a standard deviation of 3′ for the precision of a Greek measurement of the meridian altitude of the sun.

When we find a solstice time using the second definition, there are two limitations to our accuracy. One comes from the precision in measuring the meridian elevation of the sun. The other comes from the fact that the sun's orbit is not exactly a circle and hence its motion is not exactly symmetric about a solstice. I shall start by studying the second effect.

In order to study it, we need some values of the declination at times that are symmetrical about a solstice, say the summer one. I found some by reaching at random for a volume of the *American Ephemeris and Nautical Almanac* [published annually], happening to take the one for 1967. The summer solstice that year was on June 22 at $2^h\ 23^m = 2^h.38$, ephemeris time.

Table II.2 gives some declinations of the sun at times that are separated by about two weeks from the solstice in 1967. The first column gives the date and the second gives the declination at the midnight that began that date. The other columns will be explained in a moment. The fact that the declinations are calculated for midnights rather than noons clearly does not matter if we are only looking for the symmetry point of the values in Table II.2.

The declination was 22°.583 44 at the epoch July 8.0, which is the midnight beginning July 8, and it also had this value at some time between June 6.0 and June 7.0. If we use linear interpolation to find the time when the declination was 22°.583 44, the result is June 6.2370. Hence, the solstice was at the midpoint between this time and July 8.0.

This gives an estimate of June 22.1185 for the time of the solstice, which is the same as June 22 at $2^h\ 51^m$. This differs from the correct time by about 28 minutes, which is small in comparison with ancient Greek er-

TABLE II.2

SOME DECLINATIONS OF THE SUN IN 1967

Date in 1967	Tabulated Declination, *degrees*	Perturbed Declination, *degrees*	Perturbation Case
June 6	22.558 28	22.508 28	c
June 7	22.664 44	22.614 44	b
July 8	22.583 44	22.633 44	a

rors of 7 hours. Thus the lack of symmetry caused by the eccentricity of the sun's orbit is not important if we use times separated by 32 days.

Now let us estimate the effect of an error in measuring the meridian elevation of the sun, which is the way an ancient Greek astronomer would find the declination of the sun. We do not have to consider biases such as refraction, because they have a symmetrical effect upon the measurements. Hence it is legitimate to use the error level appropriate to the precision rather than to the accuracy. We decided above to use 3′ for this standard deviation.

Accordingly, I shall perturb each of the three declinations in Table II.2 in turn by 3′ (=0°.05). This yields three cases:

a. The declination on July 8 is changed to 22°.633 44, with the declinations on June 6 and 7 keeping the original values in Table II.2.
b. The declination on June 7 is changed to 22°.614 44, with the declinations on June 6 and July 8 keeping the original values in Table II.2.
c. The declination on June 6 is changed to 22°.508 28, with the declinations on June 7 and July 8 keeping the original values in Table II.2.

The perturbed values, and the letters that identify the perturbation case, appear in the third and fourth columns in Table II.2.

The results are shown in Table II.3. The first column in this table gives the identifying letter of the perturbation case. The second column gives the hour on 1967 June 22 that we find for the solstice by using the corresponding perturbed value, and the third column gives the corresponding error. Since the assumed errors in measurement are independent, the resultant error is the square root of the sum of the squares, which is 7.6 hours.

TABLE II.3

THE EFFECT OF MEASUREMENT ERROR ON FINDING THE SOLSTICE TIME

Perturbation Case	Hour of Solstice on June 22	Error, *hours*
a	8.5	6.1
b	5.4	3.0
c	5.8	3.4
Resultant Error		7.6

This is slightly larger than the estimate of 7 hours that I made by analyzing Greek data, but not disturbingly so. In particular, there are several ways to reduce the estimate of 7.6 hours. For one thing, we took the standard deviation in measuring the meridian elevation to be 3′ although Hipparchus's data indicate 2′. If we had used 2′ rather than 3′, the resultant error in Table II.3 would have been only about 5.1 hours.

We could also reduce the error by increasing the time span. At the times used in Table II.2, the declination was changing by about 6′ per day, so the measurement error of 3′ was about half the daily change. If we had used times when the change was about 9′ per day, the effect of measurement error would have been multiplied by about 2/3. However, we would have increased the effect of the orbital eccentricity, and there must be a time interval that is optimum for a given size of the measurement error. I have not tried to find the optimum.

In summary, a Greek astronomer could find the time of a solstice with an error of only about 7 hours, if his measurement error was about 3′, even though the declination of the sun changes by only about 14″ during the 24 hours following a solstice.

Hartner [1977, p. 7] finds that the accuracy of a solstice found by this method is about 2 hours if the error in measuring the meridian altitude is 1′. This is reasonably consistent with the estimate I have just made, although I believe that the postulated error of measurement is too small.

I commented above that there should be an optimum separation between the dates used. I did not try to find the optimum, but it must be greater than the separation used in Table II.2. However, there is a consideration other than accuracy which says that we must not use dates that are too far from the solstice; I have no precise definition of what "too far" is. This consideration is the need to have independent data.

To illustrate the point, let us go to the extreme of separating dates and find the time of a solstice by interpolating between the times of the equinoxes. Aside from the size of the error involved, we could not use this value to find the solar parameters because it is not an independent value. It is simply a number that has been calculated from the equinox measurements, and if we tried to use it to locate the solar apogee, we would necessarily find that it was at the solstice.

Note that two separate operations are involved in finding a solstice in the method described. In the first operation, we use interpolation to find a second time that is symmetrical to a first time. In the second operation, we estimate the time of the maximum by averaging the two symmetrically placed times. I call the second operation the "averaging method" of finding the maximum of a function.

There is no doubt that Ptolemy could use the averaging method, because he uses it to find the apogees of the orbits of Mercury and Venus, as we shall see in Sections VII.2 and VIII.2. For Venus, as it happens, he does not have to precede the averaging process by an interpolation, but he does with Mercury. In other words, in finding the apogee of Mercury, Ptolemy uses a calculation that is in every way the same as the calculation used here to find a solstice, except of course for the numbers involved.

One further point should be noted. Swerdlow and other writers seem to think that the size of the measurement error is relevant to the conclusion that Ptolemy fabricated his data. Actually it is not, as I have shown or stated several times [*CCP*, p. 92, for example]. The important point is the exact agreement, to the level of rounding used, between Ptolemy's alleged observations and Hipparchus's theory. It is unlikely that this would have happened by chance, whether the expected error in a solstice measurement was 7 hours or 7 days.

CHAPTER III

THE ORBITAL PARAMETERS OF THE MOON

1. The Lunar Motion

In the standard perturbation theory of the lunar motion, the five largest terms in the expansion of the lunar longitude λ are:

$$\begin{aligned}\lambda = L &+ 6^{\circ}.29 \sin M + 0^{\circ}.22 \sin 2M \\ &+ 1^{\circ}.27 \sin (2D - M) + 0^{\circ}.66 \sin 2D \,.\end{aligned} \qquad \text{(III.1)}$$

In this, L denotes the mean longitude of the moon and M is its mean anomaly in the modern sense. That is, it is the angle from perigee (rather than from apogee) to the position of the mean moon. D is the mean elongation, which means the mean longitude of the moon minus the mean longitude of the sun. A synodic month, which is the month of ordinary usage, is the time needed for D to increase by 360°.

With reasonable accuracy, it is correct to say that D equals 0° at a new moon and that it equals 180° at a full moon. If we substitute either of these values into Equation III.1, we get

$$\lambda = L + 5^{\circ}.02 \sin M + 0^{\circ}.22 \sin 2M \,. \qquad \text{(III.2)}$$

That is, near a new or full moon, the lunar orbit acts as if it had the eccentricity needed to make the moon deviate from its mean position by a maximum of 5°.02, if we neglect the small last term. On the other hand, D is approximately 90° at the first quarter and 270° at the last quarter. If we substitute either of these values into Equation III.1 we get

$$\lambda = L + 7^{\circ}.56 \sin M + 0^{\circ}.22 \sin 2M \,. \qquad \text{(III.3)}$$

That is, at the quarters, the eccentricity of the lunar orbit appears to be half again as large as it is at the syzygies.

This phenomenon is called the evection,[1] and explaining it was one of the triumphs of the gravitational theory. Ptolemy [*Syntaxis*, Chapter V.3] quotes an observation that Hipparchus made of the moon when it was near the last quarter and when its anomaly was near 270°. Under these conditions, Hipparchus found that the moon was 7° 40′ from the mean moon. Thus, as a phenomenon, the evection was discovered by Hipparchus unless, perhaps, someone now unknown made the discovery before him.[2]

The model that Ptolemy uses for the lunar motion is exceeded in its complexity only by his model for Mercury. The complications are introduced in order to make the model exhibit the phenomenon of evection.

2. Ptolemy's Model for the Moon

Ptolemy's model for the motion of the moon is shown in Figure III.1. The model depends upon two reference directions. One is the direction $E\Upsilon$ to the vernal equinox and the other is the direction $E\bar{S}$, where $\bar{S}$ is the position of the mean sun. I have omitted the direction $E\Upsilon$ in order to simplify the figure.

In the model, three different points rotate uniformly around the earth E at different distances. These points are $\bar{M}$, which is the mean position of the moon, and C_1 and C_2, which have no simple physical interpretation. In addition, the angle γ, which is the anomaly in the Ptolemaic sense, grows linearly with time. The angle from the equinox to the line $E\bar{S}$ is the mean longitude L_S of the sun. Ptolemy (see Section II.6 above) takes L_S to be equal to 330;45 degrees at the era of Nabonassar, and it increases, in his theory, at the rate needed to make the year equal to 365 + (1/4) − (1/300) days. The angle $\bar{M}E\bar{S}$ is the mean elongation, which was defined a moment ago. Thus the angle from the equinox to the line $E\bar{M}$ is the mean longitude L_M of the moon.

The points C_1 and C_2 lie at opposite ends of a line that rotates in the opposite sense to the line $\bar{M}E$, in such a way that the angle $\bar{S}EC_1$ always equals the lunar elongation D (angle $\bar{M}E\bar{S}$).

Point $\bar{M}$ rotates uniformly around the earth E, as I have already said, but the distance $\bar{M}E$ does not remain constant as it does so. Instead,

[1]The word evection is sometimes used to mean the phenomenon and sometimes used to mean the term 1°.27 sin $(2D - M)$ in Equation III.1. The last term in that equation is called the variation.

[2]*Pedersen* [1974, p. 266] says that the evection was "first discovered by Ptolemy." If he means the phenomenon, rather than a mechanism that produces it, I believe that this is not correct. Ptolemy clearly credits Hipparchus with the relevant observations, and I see no reason to doubt him on this point.

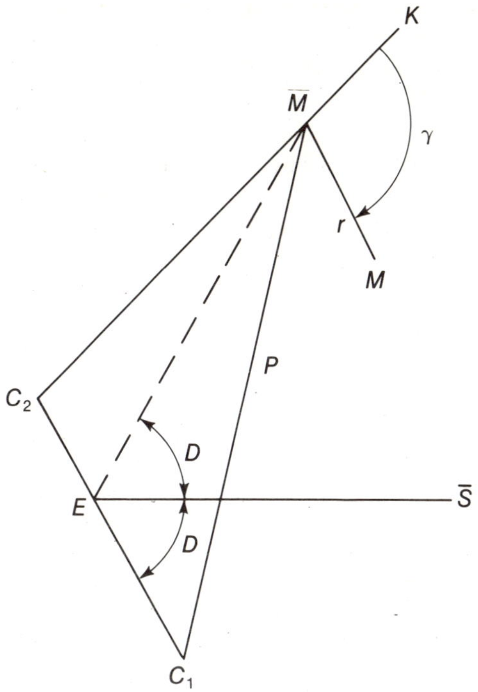

Fig. III.1—Ptolemy's model for the motion of the moon. The moon M rotates around $\overline{M}$, which is called the mean moon. $\overline{M}$ rotates uniformly around the earth E. The direction $E\overline{S}$ is the direction to the mean sun. Line C_1EC_2 rotates uniformly around E in such a way that the two angles marked D remain equal. The distance P remains constant during the motion. The anomaly, which increases at a constant rate, is measured from the line $C_2\overline{M}K$, so the rotation of M around $\overline{M}$ is not uniform. I have omitted the vernal equinox in order to simplify the figure.

the distance $\overline{M}C_1$ remains constant at the value P;[3] P is a parameter to be determined. In addition, the anomaly γ, although it grows uniformly with time, is not measured from the extension of the line $\overline{M}E$ as it is in Figure II.1. Instead, γ is measured from the line $\overline{M}K$, which is the extension of $\overline{M}C_2$, and thus M does not rotate uniformly around $\overline{M}$. γ increases in the direction shown in the figure.

[3]By the symbol P, I intend an upper case rho, not the Latin letter with the same shape.

This model does not generate a deferent curve in the usual sense. To see this, consider the appearance of the model at a time one (synodic) month after the instant for which Figure III.1 is drawn. The value of D has increased by 360°, so the triangle $\overline{M}C_1C_2$ has the same orientation with respect to the line ES that it has in the figure. In the meantime, however, the mean longitude of the sun has increased by approximately 30°, and the curve generated by the motion of $\overline{M}$ has not closed. In fact, the rates of motion of the sun and moon are probably incommensurate; if so, the curve never closes.

Let

$$\rho_1 = \text{length of } EC_1, \qquad \rho_2 = \text{length of } EC_2, \qquad \text{(III.4)}$$

for brevity. We can now enumerate the free parameters in Figure III.1. The mean longitude of the moon L_0 at a reference epoch, and the corresponding value γ_0 of the anomaly, are obviously parameters. Equation II.2 connecting L and γ does not apply here. Likewise the daily change n in L and the daily change γ' in γ are free parameters.[4] r, which is the radius MM of the epicycle (we still have an epicycle even if we do not have a deferent), is another parameter. Finally we have the three distances P, ρ_1, and ρ_2, for a total of eight parameters.

One word should be said about the distances. As long as we use the model only to get angles, we can choose one distance arbitrarily, and Ptolemy chooses to make $P + \rho_1$ equal to 60 in this case, for a reason that will appear in Section III.4. Doing this decreases the number of free parameters by one. However, determining the distance scale of Figure III.1, say the distance $P + \rho_1$, is one of the goals of Ptolemy's work, so in the general case all eight parameters are free.

We must also specify the orientation of the plane of Figure III.1 in space, and this requires the inclination i and the node Ω that were introduced in connection with Figure II.3. Further, the node is not fixed but moves uniformly around the ecliptic in Ptolemy's theory. Thus specifying the orbital plane takes three more parameters. They are the inclination i, the position Ω_0 of the node at the era of Nabonassar, and the rate of change Ω' of the node.

Finally, the apparent diameter of the moon at some standard distance is a free parameter.

Thus we have altogether twelve parameters in Ptolemy's theory of the moon. They are:

[4] I hope that it will not confuse the reader if I use the same symbols L and n here that I used in the preceding chapter for the corresponding solar parameters. When it becomes necessary to use both solar and lunar parameters in the same discussion, I shall adopt different symbols. In the meantime, it seems simpler to use the same symbols as before.

L_0, the mean longitude at the reference epoch;
n, the mean motion (daily change in L);
γ_0, the anomaly at the reference epoch;
γ', the daily change in the anomaly;
r, the epicycle radius;
P, the constant distance $\overline{MC_1}$ in Figure III.1;
ρ_1, the small distance EC_1 in Figure III.1;
ρ_2, the small distance EC_2 in Figure III.1;
i, the inclination of the lunar orbit to the ecliptic;
Ω_0, the position of the node at the reference epoch;
Ω', the rate of change of the node; and
d_M, the apparent diameter of the moon at a standard distance.

As I said in Section II.8, the distance and apparent diameter of the moon are linked to those of the sun in Ptolemy's treatment. This makes it convenient to put the discussion of these quantities into a separate chapter, except for the preliminary discussion in Section II.8. In the rest of this chapter, I shall discuss only those parameters needed to get the angular position of the moon, of which there are ten.

3. The Various Months

The sidereal month means the time interval between successive closest passages of the moon to a fixed star. (Since there is no such thing, the actual definition is somewhat more complicated, but we do not need to go into the complications.) Ptolemy does not deal with the sidereal month and we do not need to either. We can calculate it from other parameters if we should ever need it.

Ptolemy does deal with three types of month that are independent. First is the interval needed for the elongation D to increase by 360°. In simpler language, this is the interval between new moons, and it is the synodic month, the month of ordinary usage.[5] Then there is the anomalistic month, which is the interval needed for the anomaly γ to increase by 360°. Finally, there is the draconic month, which is the interval that the mean moon takes between successive passages through the position of the node.

There is also the tropical month, which is the interval that the mean moon takes between successive closest passages to the vernal equinox.

[5]Of course, the calendar month, with its irregular length, is the month that is most often used. However, I think most people know that a month, in astronomical usage, is the average interval between new moons, and this is what a lunar month means to them.

This can be derived from other data and it is not an independent parameter, although it is a useful quantity in the lunar theory.

In a passage that I mentioned in Section II.4 in connection with the sidereal year, Ptolemy gives us the basic data needed to find the various months. The passage occurs in Chapter IV.2 of the *Syntaxis*. By combining Babylonian observations with his own, Hipparchus showed that 126 007 days plus 1 hour equal 4 267 synodic months, 4 573 anomalistic months, and 4 612 sidereal months, less 7½ degrees.[6] He also found that 5 458 synodic months equal 5 923 draconic months.

As many people have pointed out, these data lead to 29;31,50,8,9 days for the length of the synodic month, but Ptolemy asserts that they lead to 29;31,50,8,20 days, "very nearly". We can accept that the values are equal, to the accuracy of the underlying data, but we ask why Ptolemy chose a value that does not come from the data cited. The answer is that 29;31,50,8,20 days is a value that Babylonian astronomers were using for the length of the month even before the time of Hipparchus [*Neugebauer*, 1955, volume 1, p. 78]. The correct value is 29;31,50,7,13 days, so Hipparchus's value is more accurate, as it should be since he had a longer time base. Yet Ptolemy accepts the older value, even though in principle it should be less accurate. Further, he says in his Chapter IV.3 that his own "observations" do not indicate any change to the value of 29;31,50,8,20 days. This is conservatism with a vengeance, and we may be sure that Ptolemy's observations were fabricated to yield the old Babylonian value. I have given explicit proof of the fabrication in Chapter VI of *CCP*.

If we multiply the adopted length of the month by the mean motion of the sun from Equation II.4a (end of Section II.3), and add 360°, we get the amount the moon moves with respect to the equinox in a month. Dividing by the length of the month gives n, the mean motion of the moon in longitude. The result is

$$n = 13;10,34,58,33,30,30 \text{ degrees/day.} \qquad \text{(III.5)}$$

Note that Ptolemy keeps seven sexagesimal positions here even though he has only five in the length of the month. This value of n comes from combining Hipparchus's length of the year with an even older value for the length of the month.

Now we may turn to the anomalistic month. Again Ptolemy does not use the data of Hipparchus that he quotes. To use Hipparchus's data, we would divide 126 007 1/24 days by 4 573 to get the anomalistic month, and we would get γ' by dividing this into 360°. Actually Ptolemy

[6]That is, the moon lacks 7½ degrees of completing 4 612 complete revolutions with respect to the fixed stars.

does not bother to calculate the anomalistic month, but goes directly to γ'. First he divides 4 573 and 4 267 by 17, getting 269 and 251; that is, 269 anomalistic months equal 251 synodic months. In 269 anomalistic months, the total change in anomaly is 269 $\times$ 360 = 96 840°, while 251 synodic months contain 7 412;10,44,51,40 days; he uses the Babylonian value of 29;31,50,8,20 in getting this result. Division then gives us

$$\gamma' = 13;3,53,56,29,38,38 \text{ degrees/day.} \qquad \text{(III.6)}$$

Hipparchus's data lead to 13;3,53,56,34 degrees per day.

However, Ptolemy does not adopt the value in Equation III.6. He says [*Syntaxis*, Chapter IV.3] that his own observations which he will present later show that the value of n in Equation III.5 is accurate but that the value of γ' in Equation III.6 needs to be decreased by 0;0,0,0,11,46,39, giving

$$\gamma' = 13;3,53,56,17,51,59 \text{ degrees/day.} \qquad \text{(III.7)}$$

I shall show the origin of this value in Section III.4.

Finally we come to the draconic month. Ptolemy does not deal with the node itself directly. Instead he uses a quantity that we call ψ, the mean argument of the latitude. ψ tells us how far the mean moon is from the node. Specifically,

$$\psi = L - \Omega.$$

The draconic month is the interval in which ψ increases by 360°.

Ptolemy says that 5 923 draconic months equal 5 458 synodic months. Repeating the calculation that gave him γ' in Equation III.6, but with these new numbers, gives him

$$\psi' = 13;13,45,39,40,17,19 \text{ degrees/day.}$$

He still uses the Babylonian rather than the Hipparchan month in this calculation. Again, he says that his own observations will show that this value is not quite correct, and he finally adopts

$$\psi' = 13;13,45,39,48,56,37 \text{ degrees/day.} \qquad \text{(III.8)}$$

I shall show the origin of this value in Section III.5.

Pedersen [1974, p. 163] says that ". . . it is difficult to escape the conclusion that the period relations used by Hipparchus . . . are of Babylonian origin." In fact the ratios 251/269 and 5 458/5 923 are found in Babylonian astronomical sources [*Neugebauer*, 1955, volume

1, pp. 76 and 73, respectively]. This does not prove that Hipparchus was dependent upon the Babylonians for these ratios, however. It may mean only that it is not possible to find better ratios of integers with a time base as short as the one Hipparchus had available. Further, if Ptolemy has quoted Hipparchus correctly, the latter did not in fact use the Babylonian periods, because his value for the synodic month, which is basic to the other periods and rates, is not equal to the Babylonian value but is better than it. Actually, it seems to me, it is Ptolemy and not Hipparchus who adopted the Babylonian values. This is true in spite of the fact that Ptolemy modified the values of γ' and Ω', as shown in Equations III.7 and III.8, as I shall show later.

4. The Mean Longitude and Anomaly at the Epoch, and the Maximum Equation

I should start this section by explaining the term in the title that I have not defined before. The actual longitude of a body minus its mean longitude is called the equation of the center, which I shall denote by e_c. Thus, repeating Equation I.2 from Section I.4:

$$e_c = \lambda - L. \tag{III.9}$$

By the maximum equation E, I mean the maximum distance that the body can get from its mean position, so that E is the maximum of $|e_c|$. A look at Figure III.1 shows us that this happens when the line of sight to the body is perpendicular to the epicycle radius. When this happens, the epicycle radius is the short leg of a right triangle whose hypotenuse is the distance from the earth to the center of the epicycle. If we let R denote this distance, clearly

$$\sin E = r/R. \tag{III.10}$$

I introduce E because I believe that Ptolemy took it rather than r as one of his original parameters.

If the reader will look at Figure III.1 and imagine what happens when D is either 0° or 180°, he will see that the points M, C_1, E, and C_2 lie on a straight line on both of these occasions, wth the distance $\overline{ME}$ equal to $P + \rho_1$; we remember that ρ_1 means the length of the line EC_1. Since we have agreed to make $\overline{ME}$ equal to 60, we have

$$P + \rho_1 = 60. \tag{III.11}$$

Back of the analysis that is to follow, and indeed back of the model shown in Figure III.1, is an unspoken assumption: Ptolemy takes it for granted that conditions are symmetrical at the new moon ($D = 0°$) and at the full moon ($D = 180°$). This assumption does not need to puzzle us. It is plausible that there was a large body of observations which showed the appropriate symmetry at all phases of the moon when it was far enough from the sun to be observed. If so, it would be natural to extrapolate this symmetry to the unobservable new moon.[7]

For simplicity, Ptolemy proposes first to use only observations made at the full moon, since the new moon is not observable. He says [*Syntaxis*, Chapter IV.1] that one should not use direct measurements of the moon's position for this purpose because parallax makes these measurements incorrect. We must pause a moment to consider the meaning of parallax, which is shown in Figure III.2.

In constructing a theory of the lunar motion, we naturally want it to be independent of the geographical position of a particular observer, so we refer it to the center *E* of the earth, as we have done in Figures II.1, II.2, and III.1. However, the observer is necessarily on the surface of the earth,[8] say at the point *T*. The direction from *T* to the observed body *B* is usually not the same as the direction from *E* to *B*, and the difference between the two directions is called parallax. The parallax depends not only upon the directions *ET* and *EB* but also on the lengths of these lines. If *B* is reasonably far away, it is a good approximation to say that the parallax is inversely proportional to the distance *EB*.

Since direct measurements of the lunar position are thrown into error by parallax, we should not use them in constructing the lunar theory, Ptolemy says, and we should use only measurements of the times of lunar eclipses, which are not affected by parallax. What Ptolemy has to say on this topic is not correct. Parallax is not an error; it is a calculable correction to an observation. To be sure, we must know the distance *EB* before we can calculate the correct direction *EB* from the observed direction *TB*, and in turn we must analyze the measurement before we know the distance. This merely means that we must analyze the measurements by successive approximations, guessing at a distance to use in the first analysis, using the measurements to refine the value of the distance, and so on.

If this seems too much trouble, we can proceed in another way. As the full moon sweeps across the night sky, it always passes through a position where its parallax in longitude is zero, regardless of distance. We can base the lunar theory upon observations made only when the parallex is zero.

[7] Actually the new moon is observable when there is an observable solar eclipse, but Ptolemy does not use solar eclipses in his work.

[8] At least this was true in Greek times.

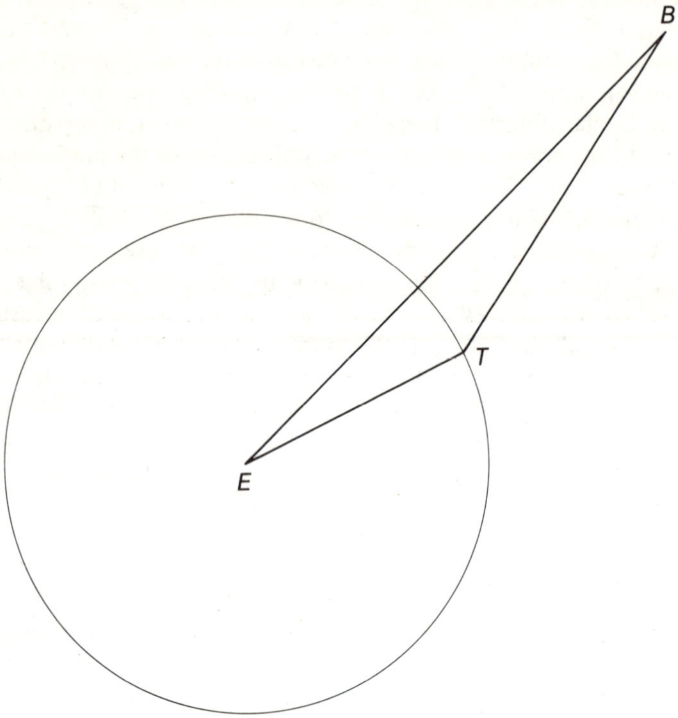

Fig. III.2—Parallax. Astronomical theories give the location of a body *B* with respect to *E*, the center of the earth, but *B* must be observed from some point *T* on the surface of the earth. The angle *EBT* is called the parallax, and it must be known in order to calculate the direction of *TB*.

Thus Ptolemy chooses to start the development of his theory by using only the measured times of lunar eclipses. Specifically, he uses, for a given eclipse, the middle time of the eclipse. This was presumably found by measuring the times when the eclipses began and ended, and taking the average.[9] Ptolemy assumes that the longitude of the moon differs from that of the sun by exactly 180° at the middle time, so he finds the longitude of the moon by calculating the longitude of the sun from his

[9]Sometimes Ptolemy uses an eclipse for which the beginning time and the magnitude were measured, but not the end time. In these cases, he calculates the duration from the magnitude, using his theory of eclipses, and adds half the duration to the beginning time.

solar theory and adding 180°. The basic assumption about the difference in longitude is not quite correct, but it is accurate enough for ancient astronomy.

I have shown [*CCP*, p. 123] that the standard deviation in the lunar longitude found from eclipses is about 15′, aside from any possible error in the solar theory used. This is about the same as the standard deviation of a direct measurement of longitude, so it is doubtful that the eclipse method is more accurate than the direct method.

Ptolemy [*Syntaxis*, Chapter IV.6] then proceeds to find three parameters of his lunar theory by using three lunar eclipses that came close together in time, specifically the eclipses of −720 March 19, −719 March 8, and −719 September 1; the observations were made in Babylon. In discussing Ptolemy's method, let me use a subscript 2 to denote the value of some quantity at the middle time of the second eclipse, that of −719 March 8. Ptolemy finds the longitude λ_2 of the moon at this time by the method I have just described. I have shown on page 114 of *CCP* that

$$\tan \lambda_2 = \frac{\sin L_2 + r \sin (L_2 - \gamma_2)}{\cos L_2 + r \cos (L_2 - \gamma_2)}, \qquad \text{(III.12)}$$

and I think the reader can easily verify this by reference to Figure III.1. In this, L_2 and γ_2 are the mean longitude and the anomaly at the middle time of the second eclipse.

Since Equation III.12 has three unknowns, namely r, L_2, and γ_2, we must use three observations and find the three parameters simultaneously. We have an equation similar to Equation III.12 for the first and third eclipses, and we relate the L's and γ's for them to L_2 and γ_2 by, for example,

$$L_1 = L_2 + n(t_1 - t_2), \; \gamma_1 = \gamma_2 + \gamma'(t_1 - t_2).$$

(Of course $t_1 - t_2$ is negative, but $t_3 - t_2$ is positive.) Over a period as short as the time between these eclipses, Ptolemy can use provisional values of n and γ' with sufficient accuracy. Thus we can express all three relations of the type of Equation III.12 in terms of the three parameters r, L_2, and γ_2.[10]

After he finds the parameters from this already ancient triad of eclipses, Ptolemy does the same thing with a triad of eclipses that he claims to have observed himself in the years 133 to 136. Altogether, in

[10]Of course Ptolemy does not use this algebraic formulation of the problem, but he uses an equivalent geometrical formulation.

Chapters IV.6 and IV.11 of the *Syntaxis*, he uses four triads of eclipses. Besides those already mentioned, one triad was observed in Babylon in the years −382 and −381 while the remaining one was observed in Alexandria in the years −200 and −199.

I have shown in Section VI.7 of *CCP* that there is at least some element of fabrication in Ptolemy's records of each of these triads, and it is likely in fact that all the data in them are fabricated. To illustrate the fabrication, let us see how Ptolemy verifies the value of n in Equation III.5 above and finds the small change in γ' used to get Equation III.7. After he analyzes the triads of eclipses in −720/−719 and in 133/136, he has values of L and γ at the times of the second eclipse in each triad, namely at specific hours on −719 March 8 and +134 October 20. The values of L are 164;44 and 29;30 degrees, respectively, and the increase in L between the two times is 224;46 degrees exclusive of complete circles. In Chapter IV.7 of the *Syntaxis*, Ptolemy calculates the change in L between the two times, using the value of n from Equation III.5. The difference between the times is more than 850 years, but Ptolemy finds agreement in L to the minute of arc! Yet, as I pointed out a moment ago, the standard deviation of a value of L found this way is about 15′. What is far worse, the error in Ptolemy's value of n is about 0.000 014 degrees per day. This makes an error of about 4°.3 over a span of 850 years.

Ptolemy also compares the values of γ at the two epochs, using the value from Equation III.6 in making the comparison. The "observed" change in γ, beyond the full revolutions, is 52;14 degrees but the calculated change is 52;31 degrees. Therefore, he says, the value in Equation III.6 is too large by enough to make an error of 17′ in a time interval that amounts to 311 783 days plus 23⅓ hours. From this he derives the difference between Equations III.6 and III.7.

Now we must ask: If Ptolemy fabricated the data using Equation III.6 for γ', why does the value of γ' deduced from the fabricated data disagree with Equation III.6, although there was no disagreement with regard to the mean longitude? We can answer this question at the same time that we answer another one: Is it in fact possible for all the data in these four triads of eclipse records to be fabricated? After all, the data agree fairly well with reality, and how could this be so if they are based upon parameters that Ptolemy invented?

I see no difficulty in assuming that all the data are fabricated. Let us start with the epicycle radius that Ptolemy finds, but instead of using this radius, let us replace it by the maximum equation E, using Equation III.10. Table III.1 shows the values of E that we derive from the four triads of eclipses. The first column in the table gives the span of years within which a triad lies and the second column gives the value of E derived from that triad. The triads appear in the order in which Ptolemy presents them, not in chronological order.

TABLE III.1

VALUES OF E FROM FOUR TRIADS OF ECLIPSES

Years	E, degrees
−720/−719	4;59
133/136	5;00
−382/−381	5;01
−200/−199	5;01

The modern value of E (the coefficient of sin M in Equation III.2) is 5°.02, which is almost the same as 5;1 degrees. Much evidence available to Ptolemy told him that the correct value was near 5°; for example, we know that Hipparchus had determined its value. My belief is that Ptolemy decided to adopt 5° exactly; we shall see other instances in which he chose a parameter simply to be some round number. It is not credible that the close agreement of the values of E happened by chance. I have estimated [*CCP*, p. 123] that the odds against the results being genuine are of the order of 64 000 to 1.

Now let us ask how he would go about fabricating eclipse data. The work of Hipparchus, and perhaps of others, would furnish him with reasonably good values of L and γ at some epoch, and he had the values of n and γ' in Equations III.5 and III.6. The trick is first to find the value of t_2, the middle time of the second eclipse in a triad. Here I think Ptolemy has to proceed by successive approximations, and he might have had genuine records of the eclipses to get him started. He guesses a value of the time, calculates the longitude of the sun from his solar theory, and adds 180° to it to get the longitude λ_2 of the moon. Then, using the same value of time, he calculates the values of L_2 and γ_2, puts them into Equation III.12, and finds another value of the longitude λ_2. Comparing the two estimates tells him how much he needs to change the time in order to make the two values of λ_2 agree. He takes $r = 60 \sin 5°$ in this process. Using the same procedure, he proceeds to find λ and the time for two other eclipses.

He would naturally keep high precision in the calculations up to this point, but now he would have to alter (by rounding or truncation) the times so that they would look like times someone might have measured. For example, he gives the times he claims to have measured himself to a precision of a quarter of an hour, and the older values are given to a precision of either a half or a third of an hour, depending upon the historical period when the "observations" were made.[11] Now, when he inverts the problems to solve for E, L_2, and γ_2 he does not get exactly the values that he started with. This explains the slight scatter in the values of E in Table III.1,[12] and it explains why he made a slight correction to the value of γ'.

The longitude λ of the moon responds 1 for 1 to a change in L, but it is rather insensitive to γ. It takes a change of more than 10′ in γ to change λ by 1′. Thus, when the rounded times are used in the inverse calculation, the rounding changes the values of L hardly at all, but it changes the values of γ by a considerable amount. Hence Ptolemy found a change to γ' but not to n.

Incidentally, even if the observations had been real, Ptolemy's change in γ' would have been illusory. Values of L found this way are uncertain by the order of 15′ and values of γ are uncertain by 150′ (2½ degrees) or more. Thus when Ptolemy found that the "observed" change in the anomaly agreed with the calculated change within 17′, this was closer than he had any right to expect.

The value of r that corresponds to taking $E = 5°$ is 5;14. Ptolemy rounds this to 5;15, which is the value of r he finally adopts. Thus r is probably the result of two roundings. First Ptolemy rounded older values of E to 5°. Then, when he found r from 60 sin E he again rounded the value, from 5;14 to 5;15 parts out of 60.

I have no specific origin for the values of L_0 and γ_0 that Ptolemy adopts, but that does not really matter for present purposes. There were undoubtedly many determinations of L_0 and γ_0 that had been made by earlier astronomers such as Hipparchus, and most of this earlier work has been lost. Thus it is not surprising that we cannot find a specific documentary source. It is possible that the values of L_0 and γ_0 derived from the very early triad of eclipses, the triad in −720 and −719, are genuine, though I think it is rather unlikely.

Let me return for a moment to the problem of finding three parameters from three values of the longitude of the moon, which Ptolemy derives from the times of three lunar eclipses. According to Ptolemy, Hipparchus had already solved this problem, and it might have been solved even earlier. When he starts his discussion of this problem, *Pedersen* [1974, pp. 173-174] writes: "Ptolemy is aware that this problem is of a very general nature. We have already dealt with a similar case when he had to determine the eccentric orbit of the Sun from observations of the two equinoctial and one of the solstitial points. Later on we shall meet it again when we determine the eccentric equant circles of the

[11]This little touch, along with much other evidence, makes me think that Ptolemy had genuine records available, and that he worded his own "records" to put them in the historically appropriate style.

[12]The standard deviation of a value of E found this way is about 15′.

planets." After noting that the problem must go back at least to Hipparchus, Pedersen goes on to say: "But I think it is Ptolemy's own merit to have given so beautiful a solution in such general terms that it can be easily *applied to similar demonstrations regardless of whether—as here—we use an epicycle or an eccentric hypothesis.*"[13]

I am afraid that I do not understand the last part of this remark. The epicycle model and the eccentric model are exactly equivalent (Section II.1), so any solution, whether it is a beautiful one or not, must apply equally to both models.

Actually I see little consistency in Ptolemy's treatment of this general problem of finding three parameters from three observations. In the solar case (Section II.5), aided by the fact that the longitudes involved are separated by 90°, Ptolemy gives an elegant method of first finding the apogee a and the eccentric e simultaneously, and then finding the anomaly in a separate calculation. In the lunar case, Ptolemy first finds the epicycle radius r (which is equivalent to e) by a rather lengthy method, and then he finds the anomaly and apogee (or the mean longitude) in a separate calculation. For the outer planets,[14] Ptolemy again finds a and e first, but he does so by an extremely long computation that is not even in closed form, as the solar and lunar solutions are.[15]

I am also afraid that Ptolemy's solution of the lunar and planetary cases seems awkward rather than beautiful to me. There is a simple and elegant method of solution that does apply without change to all the cases, but Ptolemy (and presumably Hipparchus before him) applies it only to the sun, and he does not even mention it in connection with the moon and planets. This is the method that I call the "method of the seasons" in Section II.5, and its simplicity comes from using only longitudes that are separated by multiples of 90°.

I think it is useful to give a demonstration of how one applies the method of the seasons to the moon, but I do not wish to take the space to do so at this point. Accordingly, I shall defer the demonstration to Appendix B, where I shall also apply the method to the planets.

[13]The emphasis is Pedersen's.

[14]Ptolemy manages to avoid having the problem of three observations ever arise with the inner planets.

[15]We should note that the solution of the solar problem, as well as of the lunar problem, was known to Hipparchus. Further, if Ptolemy is correct in saying [*Syntaxis*, Chapter IX.2] that Hipparchus applied a model combining the eccentric and epicycle models to the planets, Hipparchus must have solved the basic planetary problem also, although he may not have solved it in the form in which Ptolemy gives it.

5. The Argument of the Latitude

The position of the node is also derived from lunar eclipses, so Ptolemy deals with it before he goes on to consider the motion of the moon at phases well away from the full moon. Actually, as I said in connection with Equation III.8, Ptolemy uses the argument of the latitude ψ and its daily change ψ' rather than using the node directly. We get values of ψ_0, the value at some epoch, and of ψ' by considering partial eclipses of the moon.

The center of the earth's shadow lies in the plane of the ecliptic, and its diameter at the distance of the moon is bigger than the moon. Hence if the moon is close enough to one of the points where it crosses the ecliptic (its ascending and descending nodes), it spends some time completely within the shadow and is totally eclipsed. If it is far enough from a node, it misses the shadow altogether and there is no eclipse. At intermediate positions, part of it enters the shadow and there is a partial eclipse.

If the moon is north of the ecliptic when it undergoes a partial eclipse, its north side is free of the shadow and it is eclipsed on its south side. Obviously, if it is south of the ecliptic, it is eclipsed on its north side.

Now suppose that a particular fraction of the moon is shadowed when it is, say, north of the ecliptic and approaching its descending node, and that it has some specific value of its anomaly. What Ptolemy does is to search through the records of eclipses to find two in which all these conditions are the same.[16] If they are, the values of ψ must be the same on both occasions.

Ptolemy [*Syntaxis*, Chapter IV.9] finds that these conditions were satisfied in the eclipses of −490 April 25, observed in Babylon, and of 125 April 5, observed in Alexandria.[17] Both times the moon was eclipsed by 1/6 of its diameter on the south side. The time between the eclipses was slightly less than 224 609 days. If we calculate the amount that ψ changed using the Babylonian value[18] of ψ' (the value just above Equation III.8), the result is 9′ less than the change that Ptolemy finds by

[16]The purpose of the restriction on the anomaly is to ensure that the moon will be at the same distance from the earth, so that it will encounter the same size of the earth's shadow, on both occasions. Instead of requiring the same value of the anomaly, then, we really need to require only that the values be symmetrical about 180° if they are not the same.

[17]Ptolemy does not say that he made this observation himself, and it is earlier than any observation that he specifically claims to have made. It may have been made by a slightly earlier Alexandrian astronomer named Theon, whom Ptolemy quotes several times.

[18]At least it is Babylonian in the sense that it was derived using the Babylonian length of the month.

analyzing the eclipses. Thus he divides 9′ by 224 609 and adds the quotient to the Babylonian value. This gives him his final value of ψ', the value in Equation III.8.

I show on page 139 of *CCP* that the standard deviation of the change in ψ, found in this way by using two eclipses, is about 42′. Now we do not expect Ptolemy to have a formal understanding of such matters as standard deviation and statistical significance. Nonetheless his writing shows that Greek astronomers at least as far back as Hipparchus had a good understanding of the effect of observational error on the values of the parameters that they inferred.

For example, fairly near the beginning of his Chapter III.1, Ptolemy claims that he has made continuous instrumental observations of the lengths of the seasons and of the year. He finds some variability in the measured values, he says, but only "by about as much as could be attributed to the error due to the construction or position of the instruments." A little farther on he quotes Hipparchus as believing that he might have made "an error of as much as a quarter day both in observation and calculation." (In fact, a quarter day is close to the error that Hipparchus did make in his equinox and solstice data, so Hipparchus's belief was well founded.) As a final example, Ptolemy says still farther on that it is hard to measure the solstices and he prefers to base the length of the year on the equinoxes.

Thus it is reasonable to expect that Ptolemy could make a fairly good assessment of the likely error in the change in ψ, and he should have realized that he was not warranted in changing the value of ψ' on the basis of his data.

The agreement that Ptolemy found for the change in ψ was 9′, about a fifth of a standard deviation. The chance that this agreement happened by accident is about 1 part in 6. Thus the agreement alone is not sufficient grounds for concluding that there has been fabrication.[19] When I studied these eclipses in *CCP*, I could find no convincing evidence one way or the other about these eclipses, so I left them in the limbo that I called "may be genuine".[20]

However, when we consider these eclipses in the total context of the *Syntaxis*, I think we must conclude that there has been some fabrication in connection with the data. It does not seem likely that Ptolemy, on this particular occasion, made a genuine calculation with genuine data when,

19However, as I wrote on page 343 of *CCP*, we never find a moderately good probability that Ptolemy's results happened by chance. In fact, we almost always find an extremely small probability.

20The argument based upon the error in Ptolemy's values of ψ' does not help here. His error in ψ' is only about 0.000 000 55 degrees per day, and this contributes an error of only about 7′ over the time interval between the two eclipses used.

on all other occasions in the *Syntaxis*, he either fabricates the data or misstates the results of the calculations.

Thus it is probable that at least one of the eclipse records used to find ψ is fabricated; the process is exactly analogous to what we found with the anomaly. Ptolemy had to start by fabricating the time, and he then had to round it to a plausible value. A change in one or both times by about a quarter of an hour is reasonable. Thus the fabricated but now rounded data would not yield back exactly the parameters that Ptolemy started with, and in fact the discrepancy of 9′ is just about the amount the moon moves in 15 minutes.

Ptolemy uses two other eclipses to find ψ_0, the value of ψ at the era of Nabonassar. These eclipses must both come when the moon is on the same side of the ecliptic, the amount of shadowing must be the same both times, and the distance must be the same, but the moon must be near different nodes on the two occasions.[21] In the case Ptolemy uses, the moon was north of the ecliptic on both occasions. On −719 March 8 it had just passed through the ascending node and on −501 November 19 it was just approaching the descending node. On both occasions the moon was eclipsed by a fourth of its diameter. Therefore, except for the small corrections due to the equation of the center, the two positions involved were symmetrical about the point where $\psi = 90°$.

The idea is the following: Except for small corrections, which can easily be made, the value of ψ was some value, X say, on the first occasion and it was 180° − X on the second occasion. Therefore ψ changed by 180° − 2X plus some integral multiple of 360° during the time between the eclipses. However, we know ψ' and we know the time interval, so we can calculate the total change in ψ. This calculation thus yields X, which is the value of ψ at the center of the first eclipse, and from this it is trivial to find the value ψ_0 at the era of Nabonassar. The value Ptolemy finds is 354;15 degrees, but he measures this from the position of maximum latitude rather than the node (zero latitude). If ψ_0 is referred to the ascending node, Ptolemy's value is equivalent to

$$\psi_0 = 84;15 \text{ degrees.} \qquad \text{(III.13)}$$

Ptolemy does not find the lunar inclination until he is ready to find the lunar distance. We shall come to the inclination in Section III.7.

[21]As I noted in *CCP* [p. 137], I believe this is not quite correct. I think one could also use eclipses on opposite sides of the same node, but the point is a minor one.

6. The Small Distances

As we have seen, Ptolemy's lunar model shown in Figure III.1 has the two small distances EC_1 and EC_2, whose lengths will be denoted by ρ_1 and ρ_2. Ptolemy finds ρ_1 in Chapter V.3 of the *Syntaxis* and ρ_2 in Chapter V.5.

In finding ρ_1, Ptolemy starts by using an observation of the sun and moon that he claims to have made on 139 February 9 at 5¼ hours before noon in apparent time.[22] The sun appeared to be at longitude 318 5/6 degrees and the moon appeared to be at longitude 219 2/3 degrees. From the difference of the longitudes, we see that the moon was near the last quarter. Ptolemy says that the moon was in the position where no parallax correction was needed for the longitude.[23] I have analyzed this observation in detail, as well as the other observations that Ptolemy uses to find ρ_1 and ρ_2, in Sections VII.2 through VII.4 of *CCP*.

We can tell that this observation is fabricated from the solar part of it alone, without looking at the lunar part. Ptolemy calculates from his solar theory that the longitude of the sun was 318;50 degrees, agreeing to the minute with the measured value. Now the instrument that Ptolemy used for this observation was described in Section VII.1 of *CCP*. Basically it measures the declination of the sun. Then, by means of the arrangement of circles that is built into it, it functions as what we call an analogue computer in modern jargon, and its "output" is the longitude of the sun that would be calculated from the measured declination.

Thus this instrument gives the longitude of the sun with respect to the correct equinox of date, without requiring prior knowledge of where the equinox is. However, we know that Ptolemy's theory gives a longitude that is too small by more than a degree at the time of Ptolemy's observations, and the observation could not agree with his theory if the observation were genuine. Refraction and other error sources are inadequate to explain the agreement.

Now let us turn to the lunar part of the observation. In order to analyze it, we must look at the configuration of Figure III.1. When the moon is at either quarter, the points $\overline{M}$, C_1, E, and C_2 are all in line, just as they are at a syzygy, but with one important difference: The distance

22Because the sun does not move uniformly in the plane of the equator, but instead moves non-uniformly in the plane of the ecliptic, the time kept by the diurnal rotation of the sun is not uniform. When the real sun is in the meridian, we say that the time is noon, apparent time. When a point called the fictitious mean sun is in the meridian, we say that the time is noon, mean time. Ptolemy, and probably Hipparchus before him, could calculate mean time from apparent time and vice versa. See Chapter III.9 of the *Syntaxis*.

23That is, he was doing one of the things I suggested doing in the discussion of Figure III.2 in Section III.4.

$\overline{ME}$ is now equal to $P - \rho_1$ rather than $P + \rho_1$. Thus the maximum equation E is greater at the quarters than at the syzygies.

Since the moon was at the position where there is no parallax in longitude, the correct longitude of the moon was the measured value, namely 219;40 degrees, according to Ptolemy. The calculated mean longitude of the moon was 227;20 degrees, so the moon was 7;40 degrees behind its mean position. The calculated mean anomaly was 87;19 degrees, and this, still according to Ptolemy, is the value that makes the equation of the center equal to E, its maximum value. Hence, using Equation III.10, we have

$$E = 7;40 \text{ degrees}, \quad r/R = 0.133\,410. \qquad \text{(III.14)}$$

At the syzygies, on the other hand, r/R equals $\sin 5° = 0.087\,156$. Thus there are three possibilities. Either r must change, R must change, or both must change. Ptolemy elects to use a model in which R changes while r does not, although it would have been equally easy to make a model in which r alone changes or one in which both change. I shall return to this point in connection with the lunar distance in Section IV.1.

Since Ptolemy already knows the value of r, Equation III.14 immediately tells him that $R = P - \rho_1 = 39;22$, if $P + \rho_1$ has the conventional value of 60. Thus

$$P = 49;41 \text{ and } \rho_1 = 10;19. \qquad \text{(III.15)}$$

The kindest description I can think of for all this is that it is a bungled fabrication. The moon was not at the point where the parallax in longitude was zero; instead the parallax was 9′. Further, 87;19 degrees is not the anomaly at which the equation of the center reaches its maximum. The value of the anomaly for which this happens is $90° + E$, which is about 97;40 degrees. Ptolemy's error in this matter makes a further error of about 7′ in the analysis of the observation.

Actually I think we have here another instance of the principle that I called "economy of effort" in Section II.5. I introduced the principle in order to explain why Ptolemy chose to make $a' = 0$ for the sun. Consider the parallax first. Ptolemy could not calculate the parallax until he found the lunar distance, which he finds by a single observation that I shall take up in Section IV.4. By stating that the parallax was zero, Ptolemy separated the determination of E from the determination of the lunar distance, and this simplified his calculations by a considerable amount.

His best procedure, it seems to me, would have been the following: He should first observe the moon at a time that was safely before it reached the point of zero parallax in longitude, and from this observation he should calculate the time when it would reach the point in question. He should measure the position of the moon again at this calculated time. He could certainly make the calculation accurately enough that the parallax in longitude would be negligible at the time of the second measurement. In this way he would have succeeded in finding E without needing to know the lunar distance.

Ptolemy probably could not find a quarter moon near this time that occurred when the anomaly was close to the value of 97;40 degrees. The anomaly would not have had this value until the next day, but by then the moon would no longer have been close to the quarter. However, all he had to do, after handling the parallax correctly, was to use the relation that involves any value of e_c, the equation of the center, instead of the specialized relation in Equation III.10. In the triangle $\overline{MME}$ in Figure III.1 (the side ME is not drawn in), let us suppose for the moment that we know r and R (the side $\overline{ME}$). The angle between these sides is $180° - \gamma$, which is known. It was a standard and simple problem in Greek mathematics to solve a triangle of which two sides and the included angle are known. In modern notation, the solution for e_c is given by

$$\tan e_c = -r \sin \gamma / (R + r \cos \gamma). \qquad \text{(III.16)}$$

The minus sign comes from the fact that e_c is negative when $\sin \gamma$ is positive, as we see by reference to the figure. With γ and e_c known, it is almost as simple to find R in the general case as in the specialized case of Equation III.10 in either Greek or modern mathematics. However, Ptolemy used the incorrect special case instead of the correct general one.

It is clear that Ptolemy arranged for his observation of 139 February 9 to yield the value $E = 7;40$ degrees; E, rather than one of the distances r or R, was his primary quantity, just as it was at the syzygies. The question is why he chose this value, since it is not a particularly round number. The answer comes from an observation made by Hipparchus on -127 August 5. Immediately after he quotes his own observation and derives $E = 7;40$ degrees from it, but before he finds P and ρ_1, Ptolemy remarks that Hipparchus's results show the same value of E, and he presents the circumstances of the observation on -127 August 5. It was made at an hour that Ptolemy converts to 619 years, 314 days, and 17¾ hours of mean time since the era of Nabonassar. The measured longitude of the sun was 128 7/12 degrees and the measured longitude of the moon was 42 1/3 degrees.

Ptolemy does not say what instrument Hipparchus used to make the measurement, but it must have been one that functioned in about the same way as the instrument Ptolemy claims to have used. However, Ptolemy decides that Hipparchus's instrument was not aligned quite right. He calculates from his solar theory that the longitude of the sun was 128;20 degrees rather than 128;35, and he correspondingly reduces the longitude of the moon to 42;05.

We remember that Ptolemy's solar theory is really just that of Hipparchus, so this is equivalent to saying that Hipparchus's theory is more accurate than a single observation made by Hipparchus. If Hipparchus used several measurements of the lengths of the seasons when he found the solar parameters, Ptolemy is justified in correcting the observation. If Hipparchus used only one set of measurements, which seems to be the case, Ptolemy is not justified.[24] However, let us grant Ptolemy this point for the moment.

Ptolemy now says that the parallax in longitude was negligible, so that the moon's actual longitude was 42;05 degrees. The mean longitude was 34;25 degrees, so the equation of the center was 7;40. The anomaly was 257;47, which is close to the value of 262;20 degrees that makes e_c take on its maximum value, so he does not make a serious error in taking $E = 7;40$, in exact agreement with his own observation.

This observation of Hipparchus, then, is the origin of the value 7;40, but there is a bit more to the story. This is an example of what I have called "fabrication by miscalculation" [*CCP*, Section VII.4]. The parallax in longitude was not negligible. Instead, it amounted to 9′. If Ptolemy had reduced the observation correctly, he would have found $E = 7;49$ degrees rather than 7;40.

The observation made by Hipparchus disagrees with his own theory by the amount we expect for observations made in his time. From this fact, we conclude that the "raw data" quoted by Ptolemy are probably genuine.[25] He then proceeded to misstate the parallax by 9′, and arrived at the value of E that he adopted. The only reason I can see for his doing this was to save himself some trouble.

Actually the matter is more complicated than this. We remember that Ptolemy altered the lunar longitude that Hipparchus measured, and it is not likely that there was any justification for his doing so. If he had left it at the measured value, namely 42;20 degrees, the value when corrected for parallax would have been 42;29 degrees. Since the mean longitude was 34;25 degrees, the maximum equation E was 8;04 degrees. This differs appreciably from 7;40 degrees, and the difference arises from Ptolemy's "errors" in analysis.

We saw in connection with his own alleged observation that Ptolemy's reason for asserting zero parallax was to separate the observation for finding E from the measurement of the lunar distance. We also saw that Ptolemy could easily have gone to the condition of zero parallax

[24]The time intervals used were all exact multiples of 6 hours, which is not likely if they were based upon multiple observations.

[25]However, Ptolemy quotes them in a suspicious context, and I would not use them in astronomical, as opposed to historical, research.

in his fabricated observation, but that he did not take the trouble. Here, if the data attributed to Hipparchus are indeed genuine, Ptolemy did not have this option; he could not readily and correctly alter the time in order to give an "observation" made at zero parallax. Thus he would have had to analyze the data to give *E* and the lunar distance simultaneously instead of separately.

Many studies of Ptolemy praise him for his ability to separate the parameters from each other by his clever choice of observations. We now see that this is not quite correct. His observations do not give us nicely separated conditions for finding various parameters. He has separated the parameters only by making incorrect statements about the observations.

Now we turn to the small distance ρ_2. As we see from Figure III.1, the point C_2 has no effect at the syzygies or the quarters. In both cases C_2 is in line with $\overline{ME}$ and the anomaly γ is measured from the extension of this line, as it would be with the ordinary epicycle model. Also, in both these cases, *D* is a multiple of 90°. However, if *D* is far from a multiple of 90°, as it is in Figure III.1, the anomaly in Ptolemy's model is measured from the direction $\overline{MK}$, and this is far from being the extension of $\overline{ME}$.

I have no doubt that extensive observational evidence available to Ptolemy showed that measurement of γ from the extension of $\overline{ME}$ resulted in a systematic divergence between observation and calculation. In other words, the data showed that the anomaly, defined in the usual way, did not increase uniformly. By measuring the anomaly from $\overline{MK}$ rather than from the extension of $\overline{ME}$, one could reduce the divergence even if the anomaly were taken to be a linear function of time.

We can see that this must be so by considering the two largest terms in Equation III.1 (other than *L*). If we put $M = 0$ in them, the result is 1°.27 sin 2*D*, which does not vanish unless *D* is a multiple of 90°. To Ptolemy or to any astronomer who worked with the epicyclic method, it would look as if the anomaly were not zero. This in turn would mean that, while it is measured from the extension of line $\overline{EM}$ when *D* is a multiple of 90° (the quarters and syzygies), the anomaly γ must be measured from some other direction at any other time.[26]

It is perhaps to Ptolemy's credit that he recognizes this difficulty with the anomaly and invents a model that helps to solve it. If so, we must immediately take back some of the credit because of the way Ptolemy handles the model after he invents it. He uses two observations

[26]Contrary to statements by *Tannery* [1893, pp. 211ff] and other writers, the effect of C_2 does not represent the last term in Equation III.1 (the variation). The only thing that C_2 has in common with the variation is that both have their maximum effect when *D* is an odd multiple of 45°. Nothing in Ptolemy's model represents the variation at all. See *Dreyer* [1905, p. 256] or *CCP* [pp. 163ff].

made by Hipparchus on −126 May 2 and −126 July 7. The elongation D was almost exactly 315° on the first occasion and almost exactly 45° on the second. I have analyzed these observations in detail in *CCP* [Section VII.3], and I shall not repeat the analysis here. I shall only state Ptolemy's results. From the first observation he finds that ρ_2 equals 10;18 and from the second he finds that it equals 10;20. The average of these is 10;19, which is exactly the value he finds for ρ_1. Thus Ptolemy finally adopts

$$\rho_2 = 10;19. \qquad \text{(III.17)}$$

It is not credible that such close agreement of the two observations with each other, and the exact agreement of the average ρ_2 with ρ_1, happened by chance. Ptolemy must have chosen arbitrarily to make ρ_2 equal to ρ_1. He then took some observations by Hipparchus which were probably genuine in their original form,[27] and altered them so that they would yield the value for ρ_2 that he wanted. The slight scatter in the values of ρ_2 probably came from his need to round the fabricated values of the time.[28]

In summary, Ptolemy chose his parameter ρ_1 on the basis of a (probably) genuine observation by Hipparchus, but which he analyzed incorrectly. He then chose ρ_2 arbitrarily to make it equal ρ_1.

7. The Inclination of the Moon's Orbit

As Ptolemy arranges his work, he does not need the lunar inclination until Chapter V.13 of the *Syntaxis*, and he derives the value of the inclination immediately preceding this point, in Chapter V.12.

First he describes the instrument that he used to measure the inclination. He calls it an instrument for determining parallaxes, but it is really an instrument for measuring the zenith distance of a body when it is in the meridian plane. By zenith distance, I mean the angle between the zenith and the body; it is perhaps unfortunate that this has come to be

[27] It is also possible that there was no original form. That is, Ptolemy may simply have fabricated the observations *in toto*.

[28] There is an interesting facet about the observation of −126 July 7. The time in the "record" is stated to be in apparent time. When I convert this to mean time, using Ptolemy's method of making the conversion, and repeat Ptolemy's analysis, I get results that are quite different from his. However, if I assume that the time is mean time to start with, I agree closely with him. In his fabrication, Ptolemy first had to work in mean time, but he then had to convert to apparent time in order to maintain Hipparchus's known style of stating observations. I believe Ptolemy forgot to make the conversion. See page 159 of *CCP*.

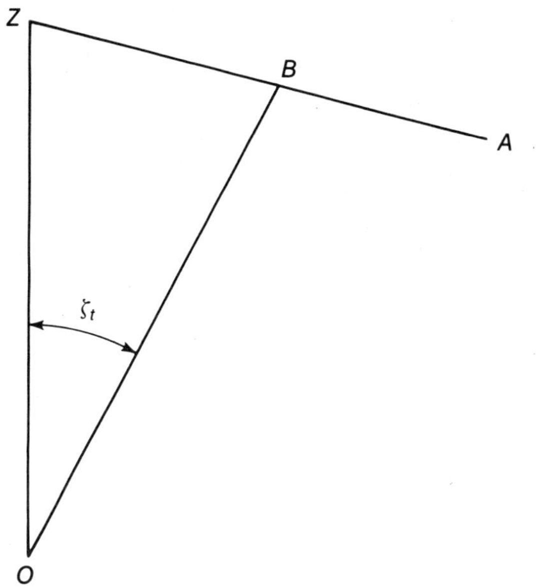

Fig. III.3—Ptolemy's instrument for measuring the zenith distance. There are hinges at points *O* and *Z*, and a pin at point *B* attached to the side *OB* slides in a groove on side *ZBA*. Sides *OZ* and *OB* are equal by construction. *OZ* is vertical while *OB* is aimed at the body to be observed. Side *ZB* is then read on a scale.

called a distance rather than an angle. The principle of the instrument is shown in Figure III.3.

The observer must arrange matters so that the plane of Figure III.3 lies in his meridian plane, and Ptolemy describes the precautions he took in doing this. In this plane, *O* is the observer's eye and *Z* is his zenith. In other words, the line *OZ* is vertical, and Ptolemy describes the precautions taken to make this so. The instrument has three arms. The arm *OZ* may be called the vertical arm. The arm *OB* is hinged to the arm *OZ*, and it is rotated until it points at the celestial object in question. We may call it the sighting arm. The third arm, which we can call the scale arm, is hinged to the vertical arm. It has a groove running its entire length, and a pin on the sighting arm slides in the scale arm at *B*.

As the name implies, the scale arm *ZBA* carries a scale, and the distance *ZB* is read off this scale. The distances *OZ* and *OB* are made to be equal, and they are known from the construction of the instrument. The zenith distance is ζ_t; the reason for the subscript will appear in a moment. We find ζ_t from

$$\text{chord}\ \zeta_t = ZB/OZ. \qquad \text{(III.18)}$$

The sighting arm carries two sights. The one at the observer's eye O is a pin hole, and the one at the other end is a larger hole. Although Ptolemy does not mention the matter, the sight with the larger hole must be able to slide along the sighting arm until the apparent diameters of the hole and of the moon are the same.[29] Only in this way can the arm be pointed accurately.

Suppose for the moment that the observer were at the equator looking at an object whose declination is zero. When the object passes through the observer's north-south plane, it is directly in the zenith, and the zenith distance ζ is zero. Now suppose that the observer moves north. Then ζ increases at the same rate as his latitude β. Now let the declination δ also increase; this decreases ζ. Thus, if β and δ are not too large, and if $\delta \leq \beta$,

$$\zeta = \beta - \delta. \qquad \text{(III.19)}$$

Equation III.19 will apply to all the cases we need to consider.

Up to now I have been trying to divert the reader's attention away from the fact that the observer is not at the center of the earth. But he is not, and the zenith distance ζ_t that he sees (the subscript denotes "topocentric") is affected by parallax. The value of ζ in Equation III.19 may be called the "geocentric zenith distance". It is the "zenith distance" an observer at the center of the earth would measure if he took the direction OZ in Figure III.3 as an artificial zenith. If Π denotes the amount of parallax, the observer's zenith distance ζ_t is

$$\zeta_t = \beta - \delta + \Pi. \qquad \text{(III.20)}$$

Thus, if one knows β and δ, and if one measures ζ_t, one can find the parallax from Equation III.20.

Suppose that the moon is in the ecliptic (celestial latitude[30] equal to zero) at the position of the summer solstice (longitude equal to 90°).

[29] In fact, the sighting arm should be just like the *dioptra*, whose accuracy Ptolemy decries. This is the instrument, described in Section II.8 above, which Ptolemy claims that he used in an attempt to measure the apparent diameter of the moon, with inaccurate results.

[30] I use "celestial latitude" for the moment to denote a coordinate measured in the heavens, because I have been talking about the observer's latitude in the same context. I omit "celestial" when there is no danger of confusion.

Then its declination equals ϵ, the obliquity of the ecliptic. Suppose, on the other hand, that it is at its position of maximum latitude, but that it is just north of the solstice point. Then its declination is $\epsilon + i$, where i is the inclination of its orbit. If we use this in Equation III.20, we have

$$\zeta_{\prime} = \beta - \epsilon - i + \Pi. \tag{III.21}$$

At Alexandria, where Ptolemy claims to have made his measurements, β is about 31° 13′. In his time ϵ was about 23° 41′. The inclination i varies slightly because of perturbations produced by the sun and planets, but its average value is about 5° 9′. Then the geocentric zenith distance ζ, from Equation III.19, is about 2° 23′. Under these conditions the parallax is small enough that it can be neglected, to the accuracy of observations that could be made in Ptolemy's time.

Ptolemy says that he measured the zenith distance of the moon "especially" when its celestial latitude was the same as i and when its longitude was 90°. He always found the zenith distance to be about 2⅛ degrees, a figure that he equates to 2° 7′ in his calculations. As his latitude is 30° 58′ (see Section II.7 above), and since ϵ equals 23° 51′ (he omits the 20″ here), Equation III.21 gives us

$$i = 5° \tag{III.22}$$

exactly, if we neglect the parallax.

It is remarkable that Ptolemy has large errors in his basic parameters β and ϵ, that he makes an error of about 16′ in measuring $\zeta_{\prime}$, and that he comes out with i equal to the round value 5°. We remember that he did a similar thing in choosing 5° for the maximum equation E at the syzygies. In other words, there can be little doubt that the zenith distance was fabricated to yield $i = 5°$, a value that he knew to be reasonably close to the results obtained by others.

There is other evidence that the result is fabricated. As I said earlier [*CCP*, p. 185], Ptolemy slips badly when he says that he "always" got the same result. For the moon to be in the position stated, both its longitude and its argument of the latitude must be 90°. This means that the longitude of the node is 0°, so that the node is at the vernal equinox.

The moon's node precesses slowly around the ecliptic and makes a complete circuit, from the equinox back to the equinox, in about 18.6 years. If he "always" got about the same result, he must have made the measurement at least three times. Conditions were right in the summer of 126 [*CCP*, p. 185] and they were not right again until the late winter of 145. Both dates are outside the range of dates on which Ptolemy claims to have made observations; these dates range from 127 March 26 to 141 February 2. Conditions were also right about the beginning of 108 and in the autumn of 163; these are far outside the accepted dates of his astronomical activity.

Ptolemy understood correctly the latitude and longitude that the moon needed in order to let him neglect the parallax when he tried to find the inclination. However, in his desire to impress us with the care that he used, he overlooked the position of the node that was implied, and the fact that he could not have made the measurement many times. Even to assume that he made it once stretches the dates involved.

We have now discussed all the parameters needed to find the latitude and longitude of the moon, and we are ready to take up its distance and apparent diameter. We shall do this in the next chapter, along with the same quantities for the sun.

CHAPTER IV

THE SIZES AND DISTANCES OF THE SUN AND MOON

1. The Variability of the Lunar Distance

The eccentricity of the lunar orbit, using "eccentricity" in its modern sense, is about 0.055, so the ratio of the greatest to the least distance to the moon is about $(1 + 0.055)/(1 - 0.055)$, which is approximately equal to 1.12.

In Ptolemy's model, which is shown in Figure III.1, the greatest distance to the moon comes when γ is zero at the time of a syzygy. When this happens, the distance is $P + \rho_1 + r$, and this equals 65;15 in the conventional system that makes $P + \rho_1$ equal to 60. The minimum distance comes when $\gamma = 180°$ at the time of either quarter, and this distance is $P - \rho_1 - r$. When we use the values of P and ρ_1 from Equations III.15, we find that the least distance is 34;7. The ratio of the greatest and least distances is 1.91 in decimal form.

If the apparent size of a body is small (a few degrees or less), the apparent size varies inversely with the distance, and Ptolemy of course knew enough geometry to know this. Thus, Ptolemy's model implies that the greatest size the moon can have at a quarter moon is almost twice the smallest size it can have at the full moon. This is manifestly not so, and Ptolemy's model is immediately shown to be seriously in error.

As I noted in Section III.6, the evection can be explained by a model in which either r or R is made to vary. Ptolemy's first choice was to vary R, and he certainly cannot be criticized for trying this first. However, this choice immediately led to a result that is contrary to fact. When Ptolemy found this out, he should have gone back to try a model that made r vary instead of R, and it is not difficult to devise such a model. If Ptolemy had done this, he would have made an important contribution to astronomy. Instead, he fabricated data to make it appear that his model was highly accurate, and in doing so he retarded the development of astronomy for more than a millenium. It was not until the fourteenth century, so far as we know, that anyone devised a lunar model in which r was made to vary rather than R [see *Neugebauer,* 1957, pp. 196ff].

In Section VII.2 we shall see another instance in which Ptolemy had to make a choice between two possibilities, with no way to know in advance which one was correct. Again he guessed wrong, and again he fabricated data to make it appear that his choice was right.

2. The Apparent Sizes of the Sun and Moon

In Chapter V.14 of the *Syntaxis,* Ptolemy describes how to make a *dioptra* and argues that it can be used only to compare diameters, but not to get their actual values. He finds, as I discussed in Section II.8, that the apparent diameter of the sun does not vary by an appreciable amount, and that the (constant) diameter of the sun is equal to the minimum diameter of the moon. In order to get the actual diameters, he argues, we must use lunar eclipses.

Before I take up his determination of the diameters, let me point out what Ptolemy has accomplished by his arguments. The direct methods of finding the lunar diameter that Ptolemy dismisses are ones that can be used at any phase, such as the quarter moons. Lunar eclipses, on the other hand, occur only at the full moon. If the direct methods were used, they would show that there is little difference between the diameter of the quarter moon and the full moon, in serious disagreement with Ptolemy's model. Hence, if he is going to argue that his model is accurate, he is forced to argue that the direct methods are inaccurate and that one should not accept their results.

Ptolemy's elaborateness about finding the lunar diameter is one of the things which convinces me that he knew about the serious error in his model. Thus, when he fabricated data about the lunar distance, he did so in order to uphold a proposition which he knew to be false.

Now let us turn to the use of eclipses to find the apparent diameter of the moon. I discussed Ptolemy's procedures in detail in Section VIII.6 of *CCP* and I shall not repeat the discussion here. I shall only summarize the results.

Ptolemy uses two observations of lunar eclipses that had been made in Babylon at times when the anomaly was essentially zero, so that the moon was at its maximum distance and its apparent diameter was the same as that of the sun, according to him. At the middle of the eclipse of −522 July 16, the northern half of the moon's diameter was eclipsed. Hence the moon was south of the ecliptic. At this time, Ptolemy calculates, the latitude of the moon was 40′ 40″ south. That is, the center of the moon was 40′ 40″ from the ecliptic. Further, since half the diameter was eclipsed, the edge of the shadow passed directly over the center of the moon. Thus the edge of the shadow was 40′ 40″ from the ecliptic, but the center of the shadow was necessarily in the ecliptic. Hence the radius ρ_U of the shadow (umbra) at this distance is

$$\rho_U = 40' \, 40''. \tag{IV.1}$$

Ptolemy will need ρ_U when he comes to find the distance to the sun.

At the center of the eclipse of −620 April 22, one fourth of the moon's diameter was eclipsed from the southern side. Ptolemy calculates that the latitude of the moon was 48′ 30″, but the radius of the shadow was 40′ 40″ on this occasion also, because the distance was the same. The difference between 48′ 30″ and 40′ 40″ is thus one fourth of the lunar diameter d_M, so that

$$d_M = 31' \, 20'' \tag{IV.2}$$

at the greatest lunar distance. This is somewhat larger than the correct value, which is about 29′ 24″. However it is close to the average solar diameter, and Ptolemy asserts that the minimum lunar diameter equals the (constant) solar one, which equals 32′ 2″. Hence the result in Equation IV.2 is close to the result that Ptolemy asserted in advance on the basis of the *dioptra*. This is a remarkable fact.

I have shown in *CCP* [p. 196] that the standard deviation of a single determination of the lunar radius made this way is about 5′ 40″, so the standard deviation to be attached to d_M in Equation IV.2 is about 11′ 20″, slightly more than a third of the value. Yet Ptolemy finds a result that, by the most generous estimate, is in error by slightly less than 2′.

This is not all that Ptolemy does with lunar eclipses in relation to the diameter of the moon. In his Chapter VI.5, he wants to know the maximum diameter that the moon can have at a syzygy[1] in order that he can construct eclipse tables. By this time, he has found the distances of the sun and moon, and thus he can calculate everything needed for eclipses from his solar and lunar models. However, he does not, and he prefers to rely on observation, saying that there is more surety in the results if one relies upon the observed phenomena![2]

Thus he uses two lunar eclipses that occurred when the anomaly was near 180°. I shall not even summarize the argument, which goes just as

[1]This is not the maximum diameter that the moon can have anywhere, but only the maximum that can occur when the moon is new or full. The absolute maximum in his model occurs at the quarters, as we saw above.

[2]The exclamation point is mine.

before. I have analyzed the argument in Section VIII.8 of *CCP*. The diameter he finds, which I shall call the maximum eclipse diameter, is

$$d_M = 35' \ 21''. \tag{IV.3}$$

The corresponding value of ρ_U is 46′ 0″. The value in Equation IV.3, like the value in Equation IV.2, is too large by almost exactly 2′. Thus Ptolemy gets remarkably consistent results, even though there is a rather good probability, given the accuracy of his method, that the second value would be smaller, not larger, than the first. I think there is little doubt that both determinations of the lunar diameter have been fabricated, although it is possible that some of the data used are genuine. In order to fabricate the result, Ptolemy needed to fabricate only one eclipse in each pair used, but we have no way to know now which data are genuine and which are fabricated.

The apparent diameter of the moon, and its variability, had been studied by Greek astronomers several centuries before Ptolemy, and reasonably accurate results had been obtained. In fact, as I noted in Section II.8, earlier results are more accurate than those which Ptolemy presents. Thus, while we cannot identify any specific documentary source for the values in Equations IV.2 and IV.3, there is no conceptual difficulty in imagining that Ptolemy fabricated the eclipse data to yield these values.

The greater difficulty comes in finding out why Ptolemy chose to make the smallest lunar diameter equal to the average solar diameter.[3] I mentioned this matter in Section II.8 above, and I shall return to it in Section IV.6.

We should note specifically that Ptolemy takes the value of d_S, the apparent diameter of the sun, to be the same as the value of d_M in Equation IV.2.

3. Parallax in the Meridian Plane

Up to this point I have had to talk about parallax only in a qualitative manner; it is now necessary to deal with it quantitatively in order to find the distances to the sun and moon. Luckily we need only the simple case of parallax in the meridian plane; we do not need to consider the problem in all three dimensions.

[3]As I have said, Ptolemy takes the solar diameter to be constant, even though he knows that the solar distance varies and hence that the solar diameter does also. When I refer to his constant solar diameter in the rest of this work, I assume that he means for this to be the average value.

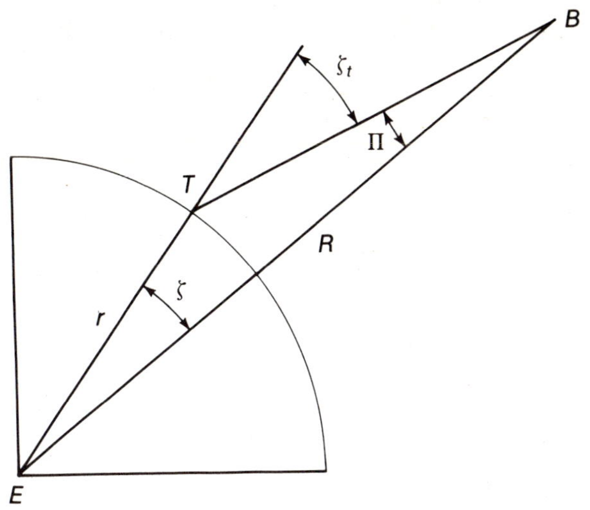

Fig. IV.1—Parallax in the meridian plane. The body being observed is at *B*, and angle ζ_t is the topocentric zenith distance measured by the observer. The angle ζ is called the geocentric zenith distance. If the ratio of *R* to *r* is known, the parallax Π can be computed if ζ is known, and conversely.

In Figure IV.1, point *E* is the center of the earth and *T* is the location of the observer. The body being observed is at *B*. The angle *TEB*, to be denoted by ζ, is what I called the geocentric zenith distance in Section III.7, while the angle ζ_t is the topocentric zenith distance that is actually observed. The parallax Π, which is the difference between ζ_t and ζ, is the angle *TBE*. Let *r* be the radius of the earth and let *R* be the distance of *B* from the center of the earth.

Since ζ_t is an exterior angle of the triangle *TEB*, the corresponding interior angle is $180° - \zeta_t$. The law of sines says that the sine of any interior angle is proportional to the length of the opposite side. If we apply this to the triangle *TEB*, using the angles ζ_t and Π, we find

$$\frac{\sin \Pi}{r} = \frac{\sin \zeta_t}{R} . \qquad \text{(IV.4)}$$

Here Ptolemy takes *r* to be unity, although he usually adopts 60 when he wants a conventional value for some distance.

In order to use Equation IV.4 to find the distance R, the observer must be able to calculate the geocentric zenith distance ζ, and he then subtracts ζ from the measured ζ_t to find Π. Then he uses Equation IV.4 to find the distance.

The parallax is a maximum when ζ_t is 90°, which puts object B on the horizon. The parallax in this particular instance is called the horizontal parallax, denoted by π, of the body B, and it no longer depends upon the position of the observer. If we replace Π by π and ζ_t by 90° in Equation IV.4,[4] and if we take r to be 1, we have

$$\sin \pi = 1/R. \qquad \text{(IV.5)}$$

Thus knowledge of the horizontal parallax of a body is equivalent to knowing its distance (if we take the earth's radius to be the unit of length), and astronomers often use the horizontal parallax rather than the distance. The reason for this is that the horizontal parallax is closer to the quantity that is observed than R is.

4. The Lunar Distance

Ptolemy finds the lunar distance in Chapter V.13 of the *Syntaxis*, immediately after he has found the inclination. He does so from a measurement of the zenith distance of the moon made when it was in the meridian. He made the observation on 135 October 1, when the moon was at the first quarter, and he found the zenith distance to be 50;55 degrees. He used the instrument described in Section III.7.

I have calculated the circumstances of the observation [*CCP*, Section VIII.5]. The value of ζ_t that Ptolemy should have measured, including the effect of refraction, is 50;14 degrees. Thus Ptolemy made an observational error of 41′ of arc. As I have pointed out, he could not have seen any part of the moon through the sighting holes of his instrument if he had really done what he said.

Ptolemy wanted to make the observation when the moon was at its maximum possible latitude and when it was simultaneously north of the winter solstice. If these conditions had been met, the declination of the moon would have been equal to $i - \epsilon$, which is equal to $-18;51,20$ degrees if we use Ptolemy's values. Then the geocentric zenith distance ζ (Equation III.19) would have been 49;49,20 degrees. However, the conditions naturally differed slightly from the ideal ones, and Ptolemy has

[4] Of course atmospheric refraction is large for a body on the horizon, and the measured zenith distance must be corrected for this effect.

to make a small correction to this, which I have verified. He finally gets 49;48 degrees for ζ.

Since the parallax is $\zeta_t - \zeta$, it comes out to be 1;07 degrees. When I use this value in Equation IV.4, along with $\zeta_t = 50;55$ degrees and $r = 1$, I find $R = 39;50$ in sexagesimal notation, while Ptolemy finds 39;45. The difference is not important, and I shall use Ptolemy's value in the rest of the discussion. The correct distance at the time of observation, incidentally, was about 60;24.

We remember that the sum $P + \rho_1$ in Figure III.1, rather than any one of the individual parameters, has been given the conventional value of 60 up to this point. The reason is that this is the average lunar distance when the moon is at its syzygies. From the distance of 39;45 earth radii found by means of the observation, Ptolemy can now show that $P + \rho_1$ equals 59 earth radii.

Greek astronomers had been studying the distance to the moon for several centuries before Ptolemy, and they knew that the average distance is about 60 earth radii (the correct value is about 60.27). I can find no specific documentation for the value 59. However, if I combine some data which Ptolemy attributes to Hipparchus with some data that Pappus[5] attributes to him, I get 59.1 earth radii [*CCP*, p. 179]. I tentatively accept this result, rounded off to 59, as the origin of Ptolemy's value, but the reasoning is tenuous and I do not urge the conclusion. In any case, I do not think the point matters for present purposes; there are many sources now lost that might have contained the value.

Thus, Ptolemy has shown that his model gives correct values for the lunar parallax at both the syzygies and the quarters, and hence that his model gives the correct variation of the lunar distance. I do not see how there can be any doubt that he fabricated his parallax measurement in order to produce these results. Further, I think he did so with full knowledge that his model was erroneous.

Ptolemy's parallax measurement does something even more remarkable. As we shall see in Section IV.6, it preserves rather accurately the ratio of the solar and lunar distances that Aristarchus had measured four centuries before Ptolemy's time, and that we discussed in Section II.8. Before we take up Ptolemy's value for the solar distance, though, it will help to have the theorem that is presented in the next section.

[5]Pappus was a Greek mathematician who flourished around the year 300. He wrote commentaries on the work of earlier Greek astronomers, including Hipparchus and Ptolemy. *Heath* [1913, pp. 341-343] gives a translation of the part of Pappus's writing that includes the data in question.

5. A Theorem on Parallaxes

For angles as small as astronomical parallaxes, the sine of an angle is proportional to the angle, to high accuracy. Thus Aristarchus's classic result, which said that the ratio of the solar to the lunar distance is 19.1,[6] can be inverted to give the ratio of the horizontal parallaxes. Since I must now talk about the sun and moon simultaneously, I shall distinguish solar and lunar quantities by the obvious subscripts *S* and *M*. Then we have:

$$\pi_M / \pi_S = 19.1. \quad \text{(IV.6)}$$

Aristarchus did not take into account the variability of the solar and lunar distances, so we do not know exactly which distances he had in mind. He probably thought that the error in taking them to be constant was small compared with the error in his measurement. For the sake of definiteness, I shall assume for the moment that Equation IV.6 applies at the mean distances.

Ptolemy has now found that the average lunar distance at the syzygies is 59 earth radii, so the horizontal parallax at this distance is $0°.971\ 16$ = 0;58,16 degrees. He could find the solar parallax immediately by using Equation IV.6, but he does not do so. Instead he uses the construction shown in Figure IV.2. The circles centered on *S* and *E* represent the sun and the earth, respectively, so that *U* is the point at the tip of the earth's shadow (umbra). Point *M* represents the center of the moon at a time when exactly half of its diameter is eclipsed. Thus the distance *MM'* is the physical radius of the umbra, and the angle *MEM'* is the apparent or angular radius ρ_U of the umbra that I introduced in Section IV.2.

If one starts from the facts that the four angles which meet at *E* add up to 180°, and that the triangles *EE'S'* and *EE'M* are right triangles, it is trivial to show that

$$\angle S'ES + \angle MEM' = \angle ES'E' + \angle EME'.$$

The first angle on the left of this equation is ρ_S, the apparent radius of the sun, and the second angle is ρ_U. The first angle on the right is the horizontal parallax π_S of the sun and the second is the horizontal parallax π_M of the moon. Hence

[6] More accurately, we should remember from Section II.8, Aristarchus said that the ratio is $1/\cos 87°$, but Aristarchus did not have an accurate value for the cosine.

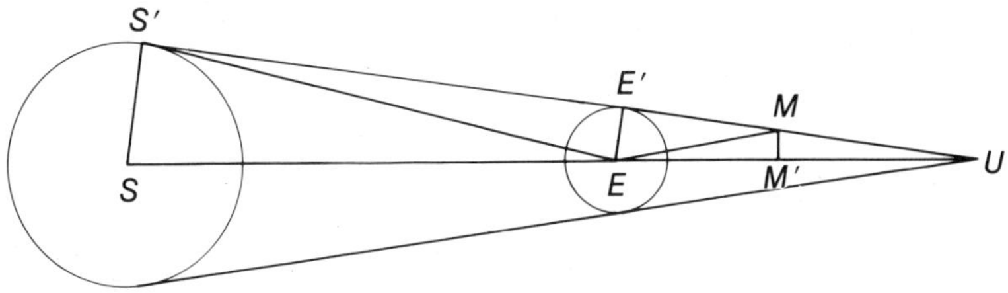

Fig. IV.2—A theorem used to find the distance from the earth *E* to the sun *S*, once the distance to the moon *M* is known. The theorem is that the sum of angles *SES'* and *MEM'* equals the sum of angles *ES'E'* and *EME'*.

$$\rho_S + \rho_U = \pi_S + \pi_M. \qquad \text{(IV.7)}$$

This theorem makes it easy to study Ptolemy's method of finding the solar distance.

6. The Solar Distance

Ptolemy finds the distance to the sun in Chapter V.15 of the *Syntaxis*. He starts by calculating the greatest value that the lunar distance can have. We saw near the beginning of Section IV.1 that this distance is 65;15 in the conventional system that makes $P + \rho_1$ equal to 60. We now have, however, that $P + \rho_1$ equals 59 earth radii. The greatest distance then turns out to be 64;10 earth radii, very nearly. The corresponding value of the horizontal parallax π_M, from Equation IV.5, is 0;53,35 degrees. As we saw in Section IV.2, Ptolemy has also shown that the umbral radius ρ_U is 0;40,40 degrees (Equation IV.1). Further, he also showed that the radius ρ_S of the sun is half of the lunar diameter given in Equation IV.2, or 0;15,40 degrees. Thus we know all the quantities in Equation IV.7 except π_S, which now turns out to be

$$\pi_S = 2'45''. \qquad \text{(IV.8)}$$

The corresponding solar distance is 1250 earth radii.

Ptolemy does not proceed in this simple fashion. Perhaps he did not know the theorem of the preceding section, or perhaps he knew it but did not think of using it. People rarely solve a problem in an easy way the first time they do it. Ptolemy starts from the values of ρ_U (angle *MEM'*) and of ρ_S (angle *SES'*) in Figure IV.2, and from the distance *EM* (the lunar distance). He then laboriously solves the various triangles until he finally gets to the distance *SE*, which is the distance he wants.

In this way, Ptolemy gets the earth-sun distance *SE* to be 1210 earth radii; the corresponding parallax is 2′ 50″. In view of the complexity of his solution, it is not surprising that this answer is in error, but the error is not serious and I shall use his values in subsequent discussion.

We remember that Ptolemy started with the lunar distance equal to 64;10 earth radii. Let us multiply this by the ratio of 19.1 found by Aristarchus (Equation IV.6).[7] The result is 1226 earth radii, with a corresponding parallax of 2′ 48″. I think "fantastic" is a reasonable adjective to apply to the agreement between Aristarchus and Ptolemy.

Thus Ptolemy has done a quite complex job of fabrication. Even when he was dealing with the lunar eclipses in order to get the umbral radius and the apparent diameters of the moon, he was arranging that he would finally come out with the ratio in Equation IV.6.

The only parameter involved in this aspect of his work that we do not have a reasonable origin for, in my opinion, and the only one that disagrees with the results of earlier astronomers, is the ratio of the apparent lunar and solar diameters. Ptolemy says that the solar diameter equals the least diameter of the moon. In contrast, earlier astronomers had found, more accurately, that the solar diameter equals the average diameter of the moon.

A possible answer comes from the work of Ptolemy called *Planetary Hypotheses*. *Planetary Hypotheses* refers specifically to the *Syntaxis* and uses some results that are developed in it. Thus *Planetary Hypotheses* was written after the *Syntaxis*, or at least after Ptolemy had planned the latter work and had fixed some of its results.

I referred to *Planetary Hypotheses* in Section I.2, where I described it as a popularized version of the *Syntaxis*. This is basically correct, but in one important respect *Planetary Hypotheses* goes beyond the *Syntaxis*. Interestingly, this is in a part of *Planetary Hypotheses* that is lost in the original Greek but that survives in both Arabic and Hebrew translations. *Goldstein* [1967] gives an English translation of the Arabic

[7]Previously I have supposed that Aristarchus's ratio was meant to apply to the mean distances, but now I have Ptolemy apply it to the maximum lunar distance. I do not believe there is an inherent contradiction. Ptolemy knew more about Aristarchus than we do, so he may have had a reason to apply the ratio to the maximum distance. More likely, I think, Ptolemy had just been working with the distance, and he did not remember to change to the average distance.

Pedersen [1974, p. 212] shows that Ptolemy could easily have gotten the solar parallax to be zero if his measured values had been slightly different from the ones he used. Actually, as I have shown in *CCP* [Section VIII.8], he would have gotten a negative parallax, which is nonsense, if he had used the data in the *Syntaxis* that lead to Equation IV.3 above rather than the data that lead to Equation IV.2. The fact that Ptolemy "happened" to use the data that lead to Equation IV.2 rather than the data that lead to Equation IV.3 by itself justifies a suspicion of fabrication.

version. When I write *Planetary Hypotheses* in the rest of this section, I shall mean specifically this particular part of the entire work.

In this work Ptolemy develops a structure of the entire universe, including all bodies believed to be a permanent part of it; this excludes such things as comets and meteors. His basic idea is the following: In the *Syntaxis* he finds that the minimum distance of the moon is 33 earth radii and that its maximum distance is 64 earth radii.[8] He associates two spheres with the moon which I shall call the inner sphere and the outer sphere. The radius of the inner sphere is 33 and the radius of the outer one is 64 (the unit of length will always be the earth radius in the rest of this section).

Ptolemy also associates inner and outer spheres with Mercury. He does not find a distance scale for Mercury in the *Syntaxis*, but only the ratio of its greatest and least distances. In *Planetary Hypotheses* he postulates that the inner sphere of Mercury has the same radius as the outer sphere of the moon, because ". . . it is not conceivable that there should be in Nature a vacuum, . . . "[9] The results of the *Syntaxis* then give him the radius of the outer sphere for Mercury. Similarly, Ptolemy says that the radius of the inner sphere of Venus has the same radius as the outer sphere of Mercury, and the work of the *Syntaxis* then gives him the radius of the outer sphere of Venus. This radius is 1079.

The sun comes next, but here there is a problem. Ptolemy has already found the distance scale of the sun in the manner just described, so he is not free to adjust the spheres of the sun to fit those of Venus. According to him, the method just described shows that the inner sphere of the sun has a radius of 1160 rather than 1079. Leaving this problem aside for the moment, Ptolemy then adjusts the inner sphere of Mars to fit the outer sphere of the sun, which has a radius, according to him, of 1260, and so on. Finally, the sphere of the fixed stars has the same radius as the outer sphere of Saturn. This radius is 19 865.

The gap between Venus and the sun is not great, and let us pretend for the moment that it is not there. Then the ratio between the outer sphere of the moon and the inner sphere of the sun is fixed by the ratios between the outer and inner spheres of Mercury and Venus. However, it is also fixed by Aristarchus's classic ratio in Equation IV.6, combined with the parameters of Ptolemy's lunar and solar models. It looks like a miracle that such disparate considerations give the same result, to high accuracy, but we may be rather sure that the miracle is of Ptolemy's making. What this result means is that Ptolemy had to work under exceedingly tight restraints in order to produce the results of both the *Syntaxis* and *Planetary Hypotheses*.

[8]These have been rounded from the values given in the *Syntaxis*.
[9]The quotation is from the translation by *Goldstein* [1967, p. 8].

But, we ask, what does all this have to do with the apparent diameters of the sun and moon, which do not seem to affect the sizes of these inner and outer spheres at all? The answer comes from Equation IV.7.[10] The *Planetary Hypotheses* and other considerations impose certain constraints on the parallaxes that appear in the right member of the equation, and thus they set limits to the value of $\rho_S + \rho_U$.

Thus, in *CCP* [p. 201], I speculated that the considerations of *Planetary Hypotheses* dictated certain values of π_S and π_M, which in turn dictated the sum of ρ_S and ρ_U. This sum in turn dictated some of the quantities involved in fabricating the eclipse data used to find ρ_U and the lunar and solar diameters. In this way, *Planetary Hypotheses* may have forced Ptolemy to make the solar diameter equal to the least diameter of the moon.

Why, the reader may also ask, do I not test this speculation by calculation? The answer is that Ptolemy does not transfer the relevant parameters accurately from the *Syntaxis* to *Planetary Hypotheses*, and he does not do his arithmetic correctly in *Planetary Hypotheses* once he has transferred them. So far I have not thought of a rigorous way to test my speculation in the face of Ptolemy's carelessness. It will take much research to tell if it is right or not, and we may never be able to tell.

As a final note, the gap that Ptolemy finds between the spheres of Venus and the sun is not real, even though he says that his observations have led him "inescapably" to the distances and ratios that he uses. Actually the gap is a result of his own carelessness. If he had carried the parameters over carefully from the *Syntaxis*, and if he had done his arithmetic correctly, he would have found [*CCP*, p. 72] that the radius of Venus's outer sphere is 1190 rather than 1079. This overlaps the inner sphere of the sun (radius 1160) instead of leaving a gap. The real situation is that small changes in some of the parameters, changes that are well within their uncertainties, would have let the outer sphere of Venus coincide exactly with the inner sphere of the sun. He was not led "inescapably" to the parameters used in *Planetary Hypotheses*, and we wonder why he felt that there was even a minor problem in joining the spheres of Venus and the sun.

In other words, after carefully arranging matters so that his cosmological scheme would come out right, why did he create this artificial problem of joining Venus and the sun? Perhaps he thought that making this minor difficulty increased the verisimilitude of his presentation, which might have looked suspicious if everything had come out exactly right, but I doubt this. In most of his work he has everything come out right at a suspicious level of precision.

[10]Although Ptolemy used a cumbersome method of solving the basic problem instead of using this simple relation, his solution must amount to the same thing.

CHAPTER V

THE STARS AND THE PRECESSION OF THE EQUINOXES

1. Ptolemy's Determinations of the Precession

Following his studies of the sun and moon, Ptolemy turns his attention to the stars and to the precession of the equinoxes. After some preliminaries in Chapter VII.1 of the *Syntaxis*, Ptolemy presents three independent ways in Chapters VII.2 and VII.3 to find the precession rate, which I shall denote by p. Then he presents a catalogue of the stars, which is divided between the last chapter of Book VII and the first chapter of Book VIII. Finally, in the rest of Book VIII, he talks about the Milky Way and some miscellaneous matters that do not concern us.

Ptolemy first determines the precession by means of an observation of the sun, moon, and Regulus (α Leonis) that he made just about sunset on 139 February 23, when the moon was at the first quarter. Just before sunset he measured the longitudes of the sun and moon. Then, just after sunset, when Regulus became visible, he measured the difference in longitude between it and the moon. He could certainly calculate the motion of the moon accurately during the interval of half an hour between the two measurements, and thus he got the longitude of Regulus. It came out to be 122;30 degrees, but Hipparchus had found it to be 119;50 degrees at a time 2⅔ centuries earlier. Thus p is exactly 1 degree per century.

It is easy to show that this observation was fabricated. Ptolemy says that he made the measurements with the instrument that I mentioned in Section III.6. It has the characteristic that it automatically gives the longitude of the sun with respect to the correct position of the equinox. But the longitude that Ptolemy claims to have measured agrees accurately with his solar theory, which is in error by more than a degree at this time in history. Thus the longitudes that Ptolemy claims to have found by observation were actually calculated.

In the next method [*Syntaxis*, Chapter VII.3], Ptolemy uses the declinations of certain stars as measured by Hipparchus and himself. He gives the necessary data for eighteen stars but he uses only six. All six stars that he uses give p close to 1 degree per century, in agreement with the first method. The interesting thing is that the twelve stars he does not use all give values of p [*CCP*, p. 221] close to the correct rate of about 1.39 degrees per century. This means that Ptolemy actually made some valid observations, but he does not use them directly. He uses them only to camouflage the fabricated measurements and to make it appear that the data used were chosen at random from a large body of data.

The presence of these genuine observations, incidentally, means that Ptolemy almost surely knew the correct rate of precession. In spite of this, he fabricated data to give a seriously wrong value.

In his last method, which appears in the same chapter of the *Syntaxis*, Ptolemy uses seven occultations of stars by the moon that had been observed at times ranging from about −280 to about +100. As I have shown in Table IX.3 of *CCP*, all seven observations were fabricated, even though they are observations that Ptolemy attributes to other astronomers. The agreement of the star positions inferred from the observations with the positions given in his star table is impossibly good. This method, too, gives p equal to 1 degree per century.

Before taking up the star catalogue, let us pause to make a short calculation. Ptolemy takes the time between Hipparchus and himself to be 2⅔ centuries. In this length of time, the total precession, if $p = 1$ degree per century, is 2° 40′. At the correct rate of precession, namely 1.39 degrees per century, the total precession is about 3° 43′. Thus Ptolemy makes an error of slightly more than a degree in this matter. We have also seen that the error in Ptolemy's theory of the sun is slightly more than a degree, almost the same as the error in the total precession. I shall study the significance of this point in Section V.3.

2. The Star Catalogue

In Chapter VII.4 of the *Syntaxis*, Ptolemy gives some details about the way he measured the coordinates of the stars given in his star catalogue. He first determined the positions of a set of bright stars by measuring their separation from the moon. He does not say so explicitly, but he may have found the position of the moon by calculation and found the position of a bright star by adding the measured separation to this calculated position.[1]

[1] In my opinion, this conventional reading is not certain. See Section V.4.

He found the positions of the other stars by comparing them with a star in this special set. He says that he measured the positions of as many stars as he could, down to those of the sixth magnitude.

It has been known for a long time that there is a bias in the longitudes in the star catalogue that makes them too small on the average; this bias is slightly more than 1°. Thus, many people have argued in this way that Ptolemy did not make the observations for the star catalogue himself: We have good evidence that Hipparchus prepared a star catalogue for the epoch −128 and that he made the measurements for it on Rhodes. This epoch is almost exactly 2⅔ centuries before the epoch of Ptolemy's catalogue. Thus, if Ptolemy had simply appropriated Hipparchus's catalogue, while adding 2° 40′ to the longitudes in order to account for the precession, his longitudes would be too small by slightly more than 1°. Hence this is what he did. I shall call this argument A.

As it happens, argument A is not valid, because there is another argument that leads to the same result, as I have pointed out in *CCP* [p. 237]; I shall call this argument B. This argument starts from the fact that the longitudes given by Ptolemy's solar theory are too small by slightly more than 1° for epochs in his own time. Further, he derived the lunar longitude from the solar longitude, so the same bias appears in his lunar theory. Thus, if he measured the stellar coordinates by referring the stars to the calculated position of the moon, the stellar longitudes would be too small by slightly more than 1° on the average.

Since both arguments lead to the same result, we cannot conclude from argument A that Ptolemy did not make the measurements. He would have made the same error in the longitudes if he had measured them by using the calculated position of the moon. In order to settle the origin of the catalogue, we must go to arguments that do not depend upon the bias in longitude. There are three such arguments, none more than a few years old, and all point to Hipparchus as the true author of the star catalogue in the *Syntaxis*.

The first argument concerns the fractions of degrees that appear in the catalogue, paying no attention to the integral number of degrees. The fractions are not recorded at uniform tabular intervals (Section I.4), and this means that different fractions should occur with different frequencies. We can derive a theoretical frequency distribution, and we find [*CCP*, Section IX.7] that the fractions in the latitude column of the catalogue agree with it quite well. The longitudes, on the other hand, do not agree, and they are definitely not a set of observed values.

However, if we take a set of observed longitudes and add a constant of the form N° 40′ to them, where N is some unknown integer, we do get the distribution that is found. Therefore this must be the way the longitudes were obtained. Now we saw above that 2° 40′ is just the amount Ptolemy would add to the longitudes observed by Hipparchus, while leaving the latitudes alone.[2]

The second argument concerns the absence of a particular type of error in the latitudes of the stars in the catalogue; this argument is due to *Rawlins* [in press]. Suppose that Ptolemy did make the measurements for the star catalogue in the way he claims. Then the bias in the stellar longitudes ultimately goes back to the bias in Ptolemy's (Hipparchus's, really) theory of the sun, which was about 1° 6′ (66′) in Ptolemy's time.

Consider, for example, the observation of the sun and Regulus that was described a few pages back. Suppose that Ptolemy did indeed align his astrolabe so that the longitude of the sun read 333°, as he claims that he did; this longitude is too small by about 66′. Now the astrolabe, like the ordinary terrestrial globe, rotates about the pole of the equator and not the pole of the ecliptic. If Ptolemy forced the astrolabe to the point where it read the wrong longitude, he necessarily forced the plane which represented the ecliptic in his instrument to make some angle with the true ecliptic. This angle is 66′ tan ϵ, which equals 29′. This in turn means that the latitudes would have a systematic error equal to 29′ cos λ, where λ is the longitude.

When Rawlins tests for an error in the latitudes of the form $E \cos \lambda$, the value he finds for E is much less in magnitude than 29′; it is in fact not significantly different from zero. Therefore the measurements for the star catalogue could not have been made in Ptolemy's time. They must have been made when the bias in the solar theory was essentially zero, and this almost by definition is the time of Hipparchus.

The third argument is one that has existed for a long time as a qualitative argument, but which *Rawlins* [in press] has made quantitative. Alexandria, where Ptolemy lived, is 5° farther south than Rhodes, where Hipparchus observed. Thus Ptolemy could see 5° farther south in the heavens than Hipparchus could. If Ptolemy indeed observed as many stars as he could, the star catalogue should include many stars that were invisible at Rhodes, but it does not. Every star in it was visible at Rhodes in the time of Hipparchus.

Specifically, if the observer is at latitude ϕ, he can see all the stars whose declination is algebraically greater than $\phi - 90°$, neglecting refraction and a few other small effects. Thus Ptolemy could see all stars whose declinations were greater than about −59°, while Hipparchus could see only those whose declinations were greater than about −54°.

I say the stars whose declinations "were" rather than "are" because the declination of a star is a function of time. The declination of a star

[2]The latitudes are not affected by precession. They do change slowly because of the rotation of the ecliptic, but the discovery of this effect lay centuries in the future.

changes slowly because of its proper motion, but the principal cause of the change is the precession of the equinoxes. Since the declination is a function of time, the most southerly stars that an observer (in the north temperate zone) can see depend upon both his latitude and the time when he observes. Thus we can infer both the latitude of the observer and the time when he observed by studying the declinations of the most southerly stars that he observed, paying due attention to refraction and the dimming of stars near the horizon because of atmospheric extinction.

Proceeding in this way, Rawlins finds the values:

$$\text{latitude} = 36°.2 + 0°.4, \quad \text{time} = -180 \pm 110 \text{ years.} \qquad \text{(V.1)}$$

If Hipparchus were the observer, the observer's latitude was about 36° and the time of the observation was about −130 or −140. These values agree excellently with Equations V.1. The values for Ptolemy, on the other hand, disagree seriously. Ptolemy's latitude was about 31°.2 and his time was about 135 years. The probability that the values in Equations V.1 could have happened by chance if Ptolemy were the observer has been estimated by Rawlins as about 10^{-13}.

It seems to me that this last argument proves something different from the first two. The last argument proves that Hipparchus, not Ptolemy, selected the stars that appear in the catalogue. If we had this argument alone, it would be conceivable that Ptolemy misrepresented only this matter of selection, and that he did actually measure the coordinates found in the catalogue. The first two arguments prove independently that he did not do this either. Hipparchus selected the stars by using all the ones he could see from Rhodes, he measured their coordinates, and he put them into a star catalogue.

In reference to the star catalogue, *Neugebauer* [1975, p. 1087] writes: ". . . there is no basis whatsoever for the assumption that a catalogue of this type existed before Ptolemy." I cannot understand this statement. The three arguments just summarized show, beyond a doubt in my opinion, that Ptolemy appropriated an earlier catalogue, and the evidence is overwhelming that the catalogue was Hipparchus's.[3] To be sure, the first two arguments did not exist when Neugebauer wrote this, but the third argument had long existed in its qualitative form. Even in its qualitative form, this argument demands the existence of an earlier catalogue which determined the selection of the stars found in the *Syntaxis*.

[3]*Gingerich* [1981] agrees that Ptolemy, without acknowledgment, probably "appropriated an existing catalogue, presumably derived from Hipparchus, . . ." rather than starting "from scratch" as he claims to do in the *Syntaxis*. "In a similar way," he writes, "theoretical astronomers today write general textbooks in which they describe how telescopes work and how observations are made even if they have never made observations themselves."

Further, there is ancient evidence for an earlier catalogue like the one in the *Syntaxis*. *Peters and Knobel* [1915] and many other writers cite a passage from Pliny (the Elder, whose dates are 23 to 79) which says, in the translation given by Peters and Knobel: "And the same individual[4] attempted what might seem presumptuous in a deity, viz.: to number the stars for posterity and to express their relations by appropriate names; having previously devised instruments by which he might mark the places and the magnitudes of each individual star." Now Pliny does not always understand his sources, and we should not take this by itself as absolute proof that Hipparchus prepared a star catalogue. Still, Pliny could hardly have written this unless a catalogue with the cited information already existed in his time, which is a few generations before Ptolemy. Since the information that Pliny mentions is exactly the information in the catalogue that Ptolemy claims as his own, it is obvious that "a catalogue of this type existed before Ptolemy."

3. The Origin of Ptolemy's Precession Rate

The preceding sections have been background to asking the question: What is the origin of the rate of precession that Ptolemy adopts, which is exactly 1 degree per century?

First let us note something about the third method that Ptolemy uses to find the precession rate p. This method uses occultations of stars by the moon. Suppose for the moment that the observations are genuine. Then, by a slight variation of the argument that I called argument B in the preceding section, we find that the analysis of these observations would indeed lead to $p = 1$ degree per century. This fact, some have argued, is the origin of Ptolemy's value.

However, the evidence points unmistakably to the conclusion that the observations are not genuine. Instead of being genuine observations that lead to the value $p = 1$ degree per century, they are observations that have been fabricated using this value.

Further, the other two methods that Ptolemy uses are not affected by the error in his solar theory. The second way, which involves measured star declinations, is obviously independent of that theory. The first method, in which the position of Regulus was referred to that of the sun, would depend upon the solar theory if the position of the sun were taken from calculation. However, Ptolemy says explicitly that he measured the solar position, and he describes how he did it. Further, as I pointed out above, his method of measurement would have given him the

[4]This refers to Hipparchus.

solar position with respect to the correct equinox, regardless of where he thought the equinox was. Thus it should have led to the correct precession, within observational error.

In summary, if the observations involved had all been genuine, Ptolemy's first two methods would have given him the correct value of p, about 1.39 degrees per century, and the third method would have given him 1 degree per century. However, all three methods depend upon data that have been fabricated using the value $p = 1$ (in the future I shall omit the units, which will always be degrees per century), so none of the methods can be the origin of Ptolmey's value for this parameter.

Now we ask: What sources of genuine data could have provided Ptolemy with a value of p, and what value would these data have given him? There are several ways he could have gotten a value of p from real data.

a. Ptolemy says in Chapter VII.2 that Hipparchus measured the longitude of Spica (α Virginis) and compared his measurement with one made by Timocharis about 1½ centuries earlier. Hipparchus found that it was 174° while Timocharis found 172°. This makes $p = 1.33$.
b. We remember that Ptolemy quoted eighteen star declinations that he claims to have measured, and that twelve of these, the ones he does not use to find p, are genuine. Hence he could have calculated p by combining his genuine observations with those made by Hipparchus, just as he did with the fabricated observations. I have made the necessary calculations [*CCP*, pp. 221ff],[5] and I find $p = 1.47$.
c. For these same eighteen stars, Ptolemy also quotes the measurements of their declinations that Timocharis or Aristyllus[6] had made about 150 years before Hipparchus, but he does not use them. He could have found p by combining these data with those of Hipparchus. When I do so, I find [*CCP*, p. 221] $p = 1.29$.
d. Ptolemy [*Syntaxis*, Chapter III.1] says that Hipparchus found that the tropical year is less than 365¼ days while the sidereal year is more. He implies that it was this discovery which led Hipparchus to the idea that the equinoxes are precessing. Now we can calculate p from the difference between the two kinds of year; Ptolemy might have felt that this was the most reliable and important way to find p. We saw in Section II.4 that the sidereal year, as measured by Hipparchus, is $365^d.259\ 859$, while his

[5] *Pannekoek* [1955] is the first person to make these calculations, so far as I know.
[6] These are two early Alexandrian astronomers.

> tropical year is $365^d.246\ 667$. The difference is $0^d.013\ 192$, so the amount the sun moves in this length of time is the amount the equinox moves in a year. The sun moves about 0.985 65 degree per day, so the precession is about 0.0130 degree per year or 1.30 degrees per century.

Thus all the sources of information available to Ptolemy, or at least all those known to us, lead to values of p that are fairly close to the correct value, and that are certainly far larger than 1. In fact, I have thought of no way to calculate p from known ancient data that leads to a value anywhere close to 1.

Since Ptolemy adopted the value 1, I believe that he did so simply because it is a round number, and that he probably knew when he adopted it that it did not agree with the data. The only possible justification for this value that I can find is a quotation from Hipparchus that Ptolemy gives near the end of his Chapter VII.2. Here, according to Ptolemy, Hipparchus says that if the equinoxes shifted not less than $0°.01$ in a year, then they shift not less than 3° in 300 years.

This is merely a multiplication and it is not a statement about the value of p. Still, if it is a valid quotation, it may give some clue about what Hipparchus had in mind. It may be that he felt that the data available to him did not give the value of p with surety, so, for safety, he said only that the value was at least 1 degree per century. If this is so, Ptolemy had two reasons for adopting this value: It is a round number, and it is a number that Hipparchus mentioned even if he never advocated it.

As a final matter, let us return to the arguments that I called A and B in the preceding section. The fact that they lead to the same bias in the star catalogue is what suggested to me that I calculate p from the sidereal year. My first thought was that the agreement would be explained if p calculated in this way turned out to be 1. However, a little thought shows that this is not so. The bias in the solar and lunar longitudes depends only upon the error in Hipparchus's length of the tropical year, and it is independent of the length he found for the sidereal year. The bias in the stellar longitudes depends only upon the fact that Ptolemy arbitrarily chose to make $p = 1$, and this choice is also independent of the length of the sidereal year. The agreement of arguments A and B is merely a coincidence.

It is also a coincidence that the value of p found from the two kinds of year is fairly accurate. The correct length of the tropical year in Hipparchus's time was about $365^d.242\ 077$ and the correct length of the sidereal year was about $365^d.256\ 358$. Hipparchus's tropical year was too long by about $0^d.004\ 590$ and his sidereal year was too long by about

$0^d.003\,501$. The difference between the two errors is about $0^d.001$, which makes an error in p of about 0.1.

It must be an accident that the two errors even have the same sign, because we make different types of measurement in finding the different kinds of year. In order to see how sensitive the value of p is to an error in one of the years, let us suppose that Hipparchus had not made any error in the sidereal year, so that he found it to be $365^d.256\,358$. The value of p we find now is 0.96, a change of 0.34 from the value we first found. Thus comparing the kinds of year would not have given the Greek astronomers an accurate value of p. In view of the difficulty of measuring stellar longitudes, the most accurate method available to them was probably the measurement of stellar declinations.

4. Hipparchus's Star Catalogue and His Value of the Obliquity

As I already pointed out in Section II.7, Ptolemy in Chapter I.12 says that both Eratosthenes and Hipparchus used 23° 51′ 20″ for the obliquity of the ecliptic. If this statement is correct, it is clear that Ptolemy adopted this value because both of the earlier astronomers had used it. If this statement is not correct, we have a more complicated situation.

As I also noted in Section II.7, there is some evidence that Hipparchus did use 23° 51′ 20″ for the obliquity. From his equinox and solstice observations, it looks as if he measured the meridian zenith distance of the summer solstice, he took the equator to be 23° 51′ 20″ farther south,[7] and he never measured the zenith distance of the winter solstice at all.

We can also deduce a value of the obliquity from the coordinates in the star catalogue, which we can now take as being definitely the work of Hipparchus rather than Ptolemy. To see how we deduce the obliquity, we must consider the construction of the astrolabe in more detail than we did earlier.

The astrolabe, or armillary sphere, rotates about an axis which the user intends to be parallel to the earth's axis, and it carries a circle normal to this axis which represents the equator. Another circle represents the ecliptic. The points where the ecliptic and equatorial circles intersect represent the equinoxes, and the angle between the two circles is made to

[7]This statement does not apply to his three earliest measurements of the autumnal equinox that have been preserved; these are not compatible with his later and more extensive series of equinox measurements. I shall return to this point in a moment.

equal the value that the user chooses for the obliquity. The addition of a latitude circle then allows the user to read the latitude and longitude of a celestial body once the equinoxes on the instrument have been lined up with their counterparts in the heavens.

There are two ways to align the astrolabe, and Ptolemy [*Syntaxis*, Chapter V.1] describes both of them. The sun must provide the primary standard by which the astrolabe is aligned. When the sun is visible, the observer rotates the astrolabe until the half of the ecliptic circle that faces the sun casts its shadow symmetrically onto the other half. When this happens, the sun is in the plane of the instrument's ecliptic circle, but by definition it is also in the celestial ecliptic circle. Thus the two circles are now parallel, within the accuracy of construction, the solar parallax, and other errors.

I have made some experiments to determine the precision of aligning an astrolabe this way, using the naked eye. Since the alignment is a symmetry operation, it can be done with high precision, I believe to a precision of about a minute of arc. We should note that this process depends only upon the *latitude* of the sun, which is zero by definition to the accuracy of ancient astronomy, and hence it puts the instrument's equinoctial points into the correct alignment with the celestial ones. Thus, contrary to a statement that *Pedersen* [1974, p. 240] and others have made, the astrolabe does furnish absolute longitudes, even if the user has no previous knowledge of where the equinoxes are with respect to the stars.

At night, when the sun is not visible, a different procedure is needed. Now we need a secondary reference whose latitude and longitude are already known. We adjust the astrolabe until the reference body appears at the known position. It is then correctly aligned, and we can now read the latitude and longitude of any other point on it. The only problem is that of establishing the secondary reference.

Ptolemy says that his secondary reference was the moon, whose position was found by calculation.[8] With the aid of the moon, then, he located a set of reference stars which served as tertiary references. He finally located most of the stars by using these tertiary references.

I have shown in the first place that Ptolemy did not make the measurements at all, and that Hipparchus was the actual author of the star catalogue. I have also shown that the *calculated* position of the moon

[8]This is the statement that is found in most literature on the subject [*Neugebauer*, 1957, p. 185, for example], and I repeated it uncritically on page 237 of *CCP*. I also repeated it near the beginning of Section V.2, but with the intention of commenting further here. What Ptolemy says is that he got the longitudes of a set of reference stars by means of the moon. In the observation of Regulus described at the beginning of Section V.1, Ptolemy claims to have located Regulus with respect to the sun (and hence to the equinox) by using the moon as an auxiliary object that was visible both before and after sunset. This may be the "means" that he intends to refer to.

was not used as a secondary reference [*CCP*, p. 238], because such a use of the moon would lead to errors that are not found in the catalogue.

The method that Hipparchus probably used was the one that Ptolemy claims to have used for Regulus, in which the longitude of Regulus was measured with respect to the sun, using the moon as an auxiliary body but not as a secondary reference. In this method, Regulus, and presumably other important stars, become secondary references.

Ptolemy is often credited with inventing the astrolabe, but I have shown [*Newton*, 1979] that Hipparchus already had one. The key to the situation is the following: First let us subtract 2° 40′ from the longitudes in the *Syntaxis* in order to restore the ones Hipparchus originally measured. Then the fractions of degrees in both the latitudes and longitudes have the statistical distribution expected of observed values. This means that the coordinates were read directly and that the longitudes were not found by measuring the separation of a star from a reference body and adding the measured separation to the previously measured longitude of the reference. Hence Hipparchus already had an astrolabe. If he did not have one, he had an instrument that served the same function, and we may as well call it an astrolabe.

Now we can go back to the obliquity that Hipparchus used. If he made an error in the value of the obliquity that he built into his astrolabe, he thereby introduced certain systematic errors into the coordinates that he measured. Thus, by studying the errors in the star catalogue, we can find his value of the obliquity ϵ. Further, because of the precession, the error for a particular star (or, better, a particular region of the heavens) depends upon the time when the observations were made. Thus we can also find the epoch of the catalogue.

Rawlins [in press] is the first person to point this out, so far as I know, and he has also performed the necessary analysis. Since it would be a mistake to assume that all the stellar coordinates were measured under the same conditions, Rawlins considers three important samples: (1) 349 stars in the zodiacal constellations, (2) fourteen stars often used by Ptolemy as references, which are listed by *Pedersen* [1974, p. 252],[9] and (3) 52 stars in the constellations Ursa Minor, Draco, and Cepheus; these stars surround both the equatorial and ecliptic poles.

For the first sample, Rawlins finds:

$$\epsilon = 23° 56' \pm 1', \text{ epoch } = -136 \pm 1 \text{ year.} \qquad (V.2)$$

[9]Except for β Aurigae, the stars in this sample also occur in the first sample, but Rawlins considers this sample separately in case it was observed under special circumstances. It turns out that this was not so, and I shall discuss only the first and third samples.

For the third sample, he finds:

$$\epsilon = 23°\,41' \ \pm 4', \text{ epoch } = -134 \pm 5 \text{ years.} \qquad (V.3)$$

The second epoch is not significantly different from the first, because of the relatively large standard deviation of the second epoch.[10] However, there are other indications that the third sample was observed later than the first one, and I shall assume this without proof in the rest of the discussion.

Thus, Rawlins concludes that Hipparchus used an obliquity of 23° 55′ (rounded from the value in Equation V.2) until about −135, when he changed to 23° 40′ (rounded from the value in Equation V.3). In other words, Hipparchus determined two values of the obliquity, neither of which agrees with the value of Eratosthenes. The first value was even more inaccurate than that of Eratosthenes, but the second was rather accurate; the correct value in Hipparchus's time was about 23° 43′. I want to make two comments about accepting these as independent values actually measured by Hipparchus, within rounding error.

First, while Eratosthenes' value of 23° 51′ (I shall drop the 20′ for simplicity in the rest of this section) is significantly smaller than 23° 56′ ± 1 statistically, I am not sure that the difference is "instrumentally" significant. That is, it seems possible to me that Hipparchus intended to build his astrolabe with an obliquity of 23° 51′ but that, because of errors in construction, he ended up with an effective obliquity of about 23° 55′. However, as Rawlins points out, there are extant calculations associated with Hipparchus which use 23° 55′ rather than 23° 51′.

Second, a value somewhat greater than 23° 41′ 30″ is consistent with Equation V.3, and Hipparchus, if he had obtained such a value, would have rounded it to 23° 45′ rather than to 23° 40′. However, other calculations attributed to Hipparchus use 23° 40′. Thus we have evidence independent of the star catalogue that Hipparchus, at different stages of his career, used both of the values 23° 55′ and 23° 40′. This evidence, taken in conjunction with his solar observations, leads to difficulties, as I noted in Section II.7. Let us review his solar observations briefly.

Ptolemy [*Syntaxis*, Chapter III.1] says that Hipparchus measured the time of the summer solstice in −134. Ptolemy does not directly state the time that Hipparchus found, but he gives enough information to let us calculate the time unambiguously. If Hipparchus measured the time of the solstice, it is reasonable to assume that he measured its zenith distance also.

[10]Incidentally, the epochs in Equations V.2 and V.3 give additional evidence that Hipparchus, not Ptolemy, is the true author of the star catalogue.

Still according to the same source, Hipparchus measured the times of the autumnal equinox in −161, −158, −157, −146, −145, and −142. The first three of these are not compatible with the last three: The point that Hipparchus took to be the equinox for the first three is farther south than the point he used for the last three by about 3′ or 4′.

Finally, still according to the same source, Hipparchus measured the times of the vernal equinox in the years −145 to −140, inclusive, and −134 to −127, inclusive. All these times, as well as the times of the autumnal equinoxes and the summer solstice, are rounded to the nearest quarter day.

All the equinoxes, whether vernal or autumnal, from −146 on, are consistent with the procedure I outlined in Section II.7: Hipparchus measured the zenith distance of the summer solstice, but he did not measure the zenith distance of the other three cardinal points. Instead, he used Eratosthenes' value of the obliquity and took the equinoctial points to be 23° 51′ south of the summer solstice. He did not measure either the time or the zenith distance of the winter solstice.

However, as Rawlins points out, the autumnal equinoxes between −161 and −157 are compatible with the same procedure, except for using 23° 55′ rather than 23° 51′. This suggests that Hipparchus used 23° 55′ from his earliest observations until about −146, and that he used 23° 51′ from then till the close of his solar observations in −127. There is no indication from his solar observations that he ever used 23° 40′. Thus we have a problem in reconciling the solar and stellar observations.

We also have to ask how Hipparchus obtained the values 23° 55′ and 23° 40′. The latter value is rather accurate, and there is no difficulty in imagining that he obtained it from measurement. But if he did, why is there no evidence of it in his solar data? The former value is so inaccurate that it seems to conflict with our general impression of Hipparchus as a careful observer. How could he have measured a value so greatly in error?

I believe that the following explanation, although it is speculative in some aspects, and although the argument is weak in some aspects, is consistent with the evidence:

1. It is not plausible that Hipparchus got 23° 55′ by using a graduated circle. He may have used a gnomon.[11] to measure this value,

[11]The gnomon is a vertical rod with a horizontal scale at its base that lets one read the length of the shadow cast by the rod at noon. The shadow is divided into an umbra and a penumbra, and the penumbra has a width of about 32′. The proper point to read is the midpoint of the penumbra, but the user may not have known this. Even if he did, the shadow is not symmetrical. Thus it is easy to make a mistake of as much as 16′ in using a gnomon, whereas there should be no difficulty in achieving a precision of the order of 2′ or 3′ with a graduated circle.

or he may have simply adopted it from some earlier source that is now lost.

2. Hipparchus built, or had built, an astrolabe using 23° 55′ for the obliquity, with no means provided for changing to a different value.
3. Sometime between −157 and −146, Hipparchus realized for some reason that 23° 55′ was not a reliable value, and he decided on some unknown basis that Eratosthenes' value was more accurate.
4. During this same interval, he built a meridian circle to observe the sun. He measured the zenith distance of the summer solstice, and he took the equator to be south of this point by Eratosthenes' value of 23° 51′. However, he could not change the obliquity used in his astrolabe, so he continued to use 23° 55′ for his stellar data. Perhaps he thought that a change of only 4′ was not worth the cost (in time, money, or effort, whichever was his limiting factor) involved in getting a new astrolabe. After all, 4′ was much less than his errors of observation in observing the stars.
5. Sometime around −135, Hipparchus actually measured the obliquity for the first time and found a value that he rounded to 23° 40′. The difference between this and the old value was 15′, which was definitely too large to ignore. Hence he got a new astrolabe that incorporated the accurate value of 23° 40′.
6. However, he made no change in his method of observing the vernal equinoxes even though he knew it led to large errors. The reason arises from his primary motive in making such a long series of measurements of the vernal equinox: He was trying to see if the length of the year was variable. For this, he needed to measure the time when the sun reached a certain constant meridian zenith distance, and it did not matter whether this was the meridian zenith distance of the equinox or not. Hence he retained Eratosthenes' value of ϵ in his solar measurements.

Whether this explanation is correct or not, it seems fairly certain that there were three pre-Ptolemaic values of ϵ.[12] One was 23° 51′, due to Eratosthenes or at least used by him. Another was 23° 55′, due perhaps to Hipparchus or perhaps to some earlier and now unknown astronomer. The third was 23° 40′, which is the only accurate Greek value known, and which was probably due to Hipparchus. The question

[12]In writing this, I am ignoring the value of 24° that was used sometimes. I suspect that this was recognized at an early date as an approximation that was used only for convenience in discussion.

TABLE V.1

SOME CALCULATED VALUES FOR THE LONGEST DAY OF THE YEAR ON THE TROPIC OF CANCER

Obliquity of the Ecliptic			D[a]		
°	′	″	h	m	s
23	40	00	13	28	.36
23	49	50[b]	13	30	00
23	55	00	13	30	45

[a]The length of the longest day of the year on the Tropic of Cancer, as a function of the obliquity of the ecliptic.

[b]Because of an error in computation, Ptolemy and presumably Eratosthenes before him thought that 23° 51′ 20″ was the value that leads to $D = 13^h\ 30^m\ 00^s$.

now is why Ptolemy chose to use Eratosthenes' value, ignoring the other poor value of 23° 55′ and also Hipparchus's accurate value.

I think the answer comes from Ptolemy's well-known proclivity for round numbers. We saw in Section II.7 that Eratosthenes probably chose his value of ϵ so that it would lead to exactly 13½ hours for the calculated length D of the longest day on the Tropic of Cancer. Table V.1 shows the values of D that we calculate from each of the three values of ϵ that we have found. Only the value 23° 49′ 50″ leads to a round value of D, and Eratosthenes' value of 23° 51′ 20″ arose from an error in calculating ϵ from this round value (13½ hours). (We should remember that this error is actually reproduced in the *Syntaxis.*)

Thus, from the three values of ϵ available to him, Ptolemy chose 23° 51′ 20″ because he thought it was the only one that corresponded to a round number for the value of D. He did not even mention the existence of the other two values.

CHAPTER VI

SOME PRELIMINARIES ABOUT THE PLANETS

1. The Epicycle Model as Applied to the Planets

The epicycle model shown in Figure II.1 is "made to order" to describe the geocentric orbits of the planets, to the accuracy with which their heliocentric orbits (and that of the earth) are circular. If we identify point *P* with the sun, so that the deferent curve is the orbit of the sun around the earth, the point *B* becomes a planet, so the epicycle is the orbit of the planet around the sun.

This interpretation creates a problem with the outer planets, because it makes the epicycle larger than the deferent for them. There is nothing wrong mathematically in doing this, but the Greek astronomers, who did not advocate the heliocentric picture, had no reason to do this and so they always kept the epicycle smaller than the deferent. Thus interpreting point *P* as the sun works for the inner planets, but we need a different interpretation for the outer planets.

For an outer planet, say Mars for example, we first take *E* to be Mars rather than the earth, we still take *P* to be the sun, and we take *B* to be the earth. Then Figure II.1 gives us the Mars-centered orbit of the earth. That is, the direction from *E* to *B* is the direction from Mars to the earth. The reverse of this direction is then the position of Mars as seen from the earth. Thus we still have a simple interpretation of the epicycle picture, but we must introduce a change of 180° in longitude at an appropriate stage in using the model.

Going back to the general case, neither the planet nor the earth has a circular orbit about the sun, so neither the deferent nor the epicycle should be a circle. Ptolemy, however, continues the practice, which he started with the moon, of keeping the epicycle circular. This throws the entire burden of "saving the phenomena"[1] onto the deferent. For Venus

[1] I shall explain this phrase in Section IX.7.

and the outer planets, this works moderately well. For Venus, the epicycle is Venus's heliocentric orbit, which has the lowest eccentricity[2] of any planet, about 0.007. For the outer planets, the epicycle is the orbit of the earth around the sun, which also has a fairly low eccentricity, about 0.017. For Mercury, on the other hand, the epicycle is its heliocentric orbit as it is with Venus, but the eccentricity has the high value of about 0.206. Ptolemy's failure to let Mercury's epicycle be something other than a circle introduces large errors.

Two angular velocities or mean motions are needed in the epicycle model. The longitude of P is the mean longitude of body B. Let us denote this longitude by L and denote the corresponding mean motion by L'. The other mean motion is the mean motion in anomaly, which we denote by γ'.[3] The mean motion n of the sun is always involved in these mean motions, but the nature of the involvement depends upon whether we have an inner or an outer planet. In discussing this involvement, let n_B denote the angular velocity of point B about P, but referred to a fixed direction rather than to the line PH, which itself has the angular velocity L'. Clearly we have

$$n_B = L' + \gamma' \qquad \text{(VI.1)}$$

for any planet.

For an inner planet, we identify P with the sun, so L' is just n, the mean motion of the sun, and n_B is the heliocentric mean motion of the planet in longitude. Then we have

$$\begin{aligned} n_B &= \text{the heliocentric mean motion,} \\ L' &= n, \\ \gamma' &= n_B - n. \end{aligned} \qquad \text{(VI.2)}$$

Ptolemy knew the second of these relations. However, he did not use the concept of the heliocentric mean motion and thus the third of these relations was not known to him. To him, γ' was simply a free parameter.

For an outer planet, L' equals L_P', the heliocentric mean motion of the planet, while n_B equals n, the mean motion of the sun. Thus, using Equation VI.1,

$$\begin{aligned} L' &= L_P'\text{, the heliocentric motion,} \\ n_B &= n, \\ \gamma' &= n - L'. \end{aligned} \qquad \text{(VI.3)}$$

[2]Here I use this term in its modern sense.

[3]Since γ always increases uniformly, Ptolemy often omits "mean" in speaking of γ', and calls it simply the motion in anomaly.

Ptolemy also knew the third of these relations, but to him L' was just the mean motion of the planet about the earth. He did not use the idea that L' was also the heliocentric mean motion.

It will always seem curious that the Greek astronomers knew the third relation of Equations VI.3 but still did not adopt the heliocentric hypothesis, which explains so simply why the relation holds for all the outer planets.

2. The Mean Motions of the Planets

Before he takes up the detailed motion of any planet, Ptolemy [*Syntaxis*, Chapter IX.3] gives some data from which the various mean motions can be obtained, as well as the resulting values. First he gives a relation of the following form for each planet:

$$N \text{ cycles of anomaly take } Y \text{ years plus } \Delta \text{ days,} \qquad \text{(VI.4)}$$

in which Δ is a relatively small number. The values of N, Y, and Δ are displayed in Table VI.1. In these relations, Ptolemy explicitly says that the year is the tropical year.

In deriving the values of γ' (the motion in anomaly) from the values of N, Y, and Δ, Ptolemy takes the length of the year to be the Hipparchan value of 365;14,48 days,[4] and he gets the values listed in the fifth column of Table VI.1. When I do the calculations, I get the values listed in the last column. I did my calculations using a pocket calculator that keeps ten decimal digits, so that the sixth sexagesimal position in my calculations is somewhat uncertain. This does not matter, however, because my results differ from those of Ptolemy beginning in the fourth position in every instance.

Before he gives a value of γ', Ptolemy gives the number of degrees that is derived from N and the number of days that is derived from Y and Δ. The number of degrees is an integer and is not rounded, but the number of days is rounded to the first sexagesimal position. To illustrate what I mean, take the case of Jupiter. 71 years minus 4.9 days equals 25 927.613 33 = 25 927;36,48 days, which Ptolemy takes to be 25 927;37 days. When I repeat the calculations using the number of days that Ptolemy explicitly states, I do get his values of γ' for Saturn, Venus, and Mercury, but I get different values for Jupiter and Mars. I deduce that he intended to derive the values of γ' from the data given in Table

[4]This is the same as 365¼ less 1/300 days.

TABLE VI.1

DATA FOR FINDING THE MOTIONS OF THE PLANETS IN ANOMALY

Planet	N,[a] *cycles*	Y,[a] *years*	Δ,[a] *days*	Motion in Anomaly, *degrees per day* Ptolemy's Value	My Value
Saturn	57	59	1.750	0;57,7,43,41,43,40	0;57,7,43,39,49,17
Jupiter	65	71	−4.900	0;54,9,2,46,26,0	0;54,9,2,44,26,11
Mars	37	79	3.217[b]	0;27,41,40,19,20,58	0;27,41,40,14,30,36
Venus	5	8	−2.300	0;36,59,25,53,11,28	0;36,59,25,34,56,50
Mercury	145	46	1.033[c]	3;6,24,6,59,35,50	3;6,24,7,47,31,24

[a] For the meaning of these quantities, see relation VI.4 in the main text.
[b] This is 3 13/60 days.
[c] This is 1 1/30 days.

VI.1, but that he made three errors in arithmetic in doing so.[5] Either he did not check his calculations or, if he did, he did not change his values of γ'. This repeats a pattern we have seen before and will see again: When he makes a mistake, he rarely corrects it. I have found many errors in his work, some of them in quite important matters, and I have found one and only one instance in which he corrected an error that he had made. This error, which is studied in Section VII.6, concerns what Ptolemy calls the perigee of Mercury.

The sexagesimal positions in Ptolemy's values of γ' are illusory after the fourth, as he should have realized. To continue with Jupiter as an example, 25 927;37 in full sexagesimal notation is 7,12,7;37, and Ptolemy obtained the last position by rounding. Thus, regardless of any question about the accuracy of the basic data, Ptolemy by his own calculations limited the accuracy of γ' to four positions. However, as we see from Table VI.1, he gives seven positions for Mercury and six for the other planets.

I suppose it is possible that Ptolemy did not actually make the mistakes in arithmetic that I have mentioned. It is conceivable that he had two sources of information. One source gave the values of N, Y, and Δ while the other gave the values of γ'. Ptolemy then just pretended that the values of γ' were derived from N, Y, and Δ without checking the matter. However, this does not explain the error with the number of days

[5] In the case of Mars, I cannot even verify his number of days. I get 28 857;42,12 days while he gets 28 857;53 days. Further, I do not get his value of γ' whether I use 28 857;53, 28 857;42,12, or 28 857;42 days.

for Mars, and it seems to me that this is necessarily a numerical error. This explanation is consistent with the fact that Ptolemy's number of days for Mars does not agree either with the values of N, Y, and Δ or with the value of γ'.

At an appropriate point in his detailed study of each planet, Ptolemy again takes up the value of γ'. In each case he first gives an observation that he made himself from which he deduces a value of γ. He follows this by quoting an already ancient observation from which he also deduces a value of γ. By dividing the change in γ by the change in time, he gets a value of γ'. He then says explicitly in each case that this value of γ' agrees with his value given in Table VI.1 because he used the data in question to derive the values of N, Y, and Δ in that table. I have analyzed these observations in detail in *CCP*.[6]

It is highly unlikely that Ptolemy derived the values of N, Y, and Δ in Table VI.1 from the observations that he quotes, quite aside from any question about the validity of the observations. Before we take up the reasons for this conclusion, though, let us first look at the circumstances under which it would be natural to find γ' by using a time interval that is almost an integral number of years.

Suppose we try to find γ' by observing a planet at successive times when it takes up some specific configuration. For the sake of definiteness, let us consider successive conjunctions with the sun.[7] When we study the intervals between conjunctions, we find that they are not constant. Instead, we find that the interval varies systematically with the position that the sun has at the time of conjunction. That is, we find that the motion of the planet depends upon two parameters. One is the motion of the sun and the other is characteristic of the individual planet, as we saw in the preceding section.

Thus, if we have no adequate theory or model of the motion, we must keep up with the conjunctions until we find two for which the sun has almost the same position, if we are to get a valid value of γ'. But if the sun has almost the same position, the time interval has been almost an integral number of years. In other words, taking the time interval to be almost an integral number of years, with the change in anomaly being exactly an integral number of revolutions, is a device we use when we have no model of the motion but are already developing numerical methods of making planetary calculations.

[6]The observations are given in the following chapters in the *Syntaxis*: IX.10 for Mercury, X.4 for Venus, X.9 for Mars, XI.3 for Jupiter, and XI.7 for Saturn. The relevant sections in *CCP* are X.2, X.3, XI.3, and XI.6.

[7]Of course a planet cannot be observed when it is in conjunction, because of the brilliance of the sun. However, we can find the time of a conjunction by interpolation between positions near conjunction when the planet was visible.

This is a perfect description of the state of Babylonian astronomy, but it does not apply to Greek astronomy in the time of Ptolemy, and probably not in the time of Hipparchus (see Section IX.2 below). Ptolemy had a model of planetary motion that would have been reasonably accurate (except for Mercury) if he had determined the parameters adequately. Thus he had no need to use nearly-integral numbers of years, and in fact he did not, as we shall see in a moment. In fact, he did not use either complete revolutions of the anomaly or nearly-integral numbers of years.

This suggests that the values in Table VI.1 are Babylonian in origin. However, the Babylonians did not know of the tropical year. The year to them was the sidereal year, and their value for it was (Section IX.4 below) $365^d.260\ 638$. Thus the Babylonians cannot have directly provided the values of Y and Δ in Table VI.1. It is possible, however, that Ptolemy found Babylonian values expressed by means of the Babylonian year and that he calculated what the values of Δ would become if he used the tropical year instead.

As I said a moment ago, it is unlikely that Ptolemy derived the values in Table VI.1 from the observations that he quotes. There are at least six reasons for this conclusion:

1. His values of γ' cannot be derived from the data quoted. For Mercury, for example, the change in anomaly between the two observations quoted was [*CCP*, p. 277] 268 complete revolutions plus 246;30 degrees, and the time interval was 402 Egyptian years (of exactly 365 days) plus 283 days plus 13½ hours.[8] These figures give $\gamma' = 3.106\ 698\ 970\ 942 = 3;6,24,6,58,39,48$ degrees per day. This does not agree with the value that Ptolemy uses in his planetary tables, nor does it agree with the value I calculate in Table VI.1.
2. Ptolemy says in effect that he derived a value of γ' from the observations that he quotes, that he then calculated the values of Y and Δ which match the values of N in Table VI.1, using his value of γ' to do so, and that he then turned around and calculated γ' again from N, Y, and Δ. There was no reason to go through this process, which involves two extra stages of computation. In fact, the values of N, Y, and Δ are superfluous.
3. Giving γ' by means of N, Y, and Δ is suitable for Babylonian astronomy but not for Greek astronomy, as I have already said.

[8]Note that the change in anomaly is not close to an integral number of revolutions and that the time interval is not close to an integral number of years. Further, Ptolemy used the Egyptian year, with a number of days that was exactly known, instead of the imperfectly known tropical year.

In fact, the values of N and Y occur in known Babylonian sources.[9] I have not found the values of Δ in a Babylonian source, but I have made no particular attempt to do so.

4. The observations that Ptolemy claims to have made himself were fabricated. In each of the observations of the inner planets that he claims to have made, he measured the separation in longitude of the moon, the planet, and a star. If we determine the longitude of the moon from the observations [*CCP*, Section X.2] and if we then calculate it from Ptolemy's lunar theory, including the parallax, the values agree to the minute of arc. Yet the standard deviation of the error in Ptolemy's lunar theory, not counting the bias in longitude, is [*CCP*, p. 163] 35 minutes of arc.[10]
5. I can think of no way in which an ancient astronomer could have used the tropical year as a unit in the precise measurement of time. It seems to me the original data must have been stated in terms of days, with a precision of 1/60 of a day, and that someone converted these data into the values of N, Y, and Δ shown in Table VI.1, using Hipparchus's length of the tropical year to do so. This person might have been anyone between Hipparchus's time and Ptolemy's.
6. Immediately after he "derives" the values of γ' in his Chapter IX.3, Ptolemy gives tables of γ as a function of time for each planet in Chapter IX.4. He does this by first giving γ at the era of Nabonassar. He then gives a table of the amount that γ changes in various intervals of time. The leading part of this table gives the change at intervals of 18 Egyptian years, up to 810 such years. He also gives tables for intervals of 1 Egyptian year, of 1 Egyptian month of 30 days, of 1 day, and of 1 hour, all up to suitable multiples of the basic unit. However, the real usefulness of the tables stops at a time that is about 810 Egyptian years after the era of Nabonassar. Now 810 Egyptian years after the era of Nabonassar, which is −746 February 26, is 63 August 8.

In other words, Ptolemy's planetary tables cease to be useful at a time that was two or three generations before his own time, and the same is true of his tables of the sun and moon. I can think of no reason why an astronomer, in preparing tables of motion of all the astronomical bodies known to him, would have made them stop in his grandparents' time, which means that they were almost useless to his successors or even to his

[9]See *Newton* [1976, pp. 101ff], which gives further references to the literature.

[10]For an outer planet, the observation Ptolemy claims to have made is an observation of the planet at opposition, and the proof of fabrication is different. See Section VIII.5.

contemporaries, including himself, in the study of their own observations.

The only plausible explanation I can think of is that Ptolemy took over his tables of motion for the sun, moon, and planets, and also the various values of γ', from some source that he does not name that was at least two generations before his own time, and probably more. It is quite possible that the tables were prepared several centuries before Ptolemy. It does not seem possible to me that they were prepared in his time.

In sum, Ptolemy's values of γ' do not come from either set of data that he quotes, and we do not in fact know where they come from. It is possible that they are Babylonian in origin. The fact that we have not found an earlier source for them does not matter for the purposes of the present work.

Neugebauer [1975, p. 152] makes an odd error[11] about the values of γ'. The only one he studies is the value for Saturn. Ptolemy rounds the number of days derived for Saturn from Table VI.1 to 21 551;18, and the number of degrees is 20 520. The quotient of these numbers is 0;57,7,43,41,43,40 degrees per day; this is the value Ptolemy uses in his tables of Saturn. In Chapter XI.7 of the *Syntaxis*, Ptolemy compares an observation that he claims to have made with an observation made about 365 years earlier, and finds that the anomaly changed by 126 711;27 degrees in 133 079 ¾ days. This quotient is 0;57,7,43,41,44,18, which differs from the first value[12] in the fifth sexagesimal position. Thus Ptolemy's value of γ' does not come from the observations quoted, as I have said before.

Neugebauer says 20 520 divided by 21 551;18 is 0;57,7,43,41,33. He stops at the fifth position, presumably because his result already deviates from Ptolemy's in this position. As a result of this error in arithmetic,[13] Neugebauer says that Ptolemy's value of γ' does not come from the data in Table VI.1. This confirms, he goes on to say, that Ptolemy found his values of γ' for all the planets from the observations given in the *Syntaxis*. It is unfortunate that Neugebauer did not calculate the value of γ'

[11] I thank Dennis Rawlins (private communication) for pointing out this error to me, as well as pointing out when Ptolemy's tables stop.

[12] I did the calculations in this paragraph, including the conversion to sexagesimal form, on a large-scale computer to a precision equivalent to about 15 decimal digits, so the values given are accurate in the sixth sexagesimal position.

[13] On his page 1113 Neugebauer says that, in dealing with sources which use the sexagesimal system, it is "essential" not to convert the sexagesimal numbers into decimal form, do the operations in decimal, and then convert back to sexagesimal. This of course is not correct. As I said in the preceding footnote, I did the calculations on a large-scale computer which accepted the inputs in sexagesimal form, converted to binary for the actual division, and then converted the result back to sexagesimal. Of course, if one encounters an error in arithmetic that was done in sexagesimal form, one might get some insight into the origin of the error by repeating the arithmetic directly in sexagesimal form.

for Saturn that does result from using the observations. Had he done so, the argument that he himself uses would have shown him that the observations are not the source of γ'.

We see from Equations VI.2 that the mean motion of an inner planet in longitude is just n, the mean motion of the sun.[14] Thus Ptolemy has all the mean motions of an inner planet that he needs once he has found γ'. For the outer planets, however, he still needs L', the mean motion in longitude. He finds it from γ' by using the third of Equations VI.3, which involves the mean motion of the sun.

3. The Maximum Elongation and the Apogee of a Planet

By the elongation D of the moon or a planet in modern astronomy, we usually mean the difference between its longitude and that of the sun. Ptolemy, however, takes D for a planet to mean its actual longitude λ minus the mean longitude L_S of the sun. That is,

$$D = \lambda - L_S. \qquad \text{(VI.5)}$$

For the outer planets, D can take on any value, and Ptolemy bases much of his work with them on observations made when D is 180°. For the inner planets, on the other hand, D is limited to about 28° for Mercury and to about 48° for Venus, and observations made when D has these values play an important part in Ptolemy's theories of those bodies. The rest of this section will deal only with the inner planets.

Ptolemy rarely uses the elongation of an inner planet except in the configuration that he calls maximum elongation. If the reader will refer again to Figure II.1 for the epicycle model, he will see that the body B is at its maximum distance from the epicycle center P when the line of sight from E is tangent to the epicycle. When an inner planet is in this situation, Ptolemy says that it is at maximum elongation. He then finds the value of D from Equation VI.5, using a measured value of λ and a calculated value of L_S.[15]

The elongation as defined by Equation VI.5 can be either positive or negative, but Ptolemy does not use the negative sign. He states a value of

[14]However, Ptolemy does not identify the center of the epicycle with the sun. He just says that it moves at the same rate as the sun. In fact, in his *Planetary Hypotheses*, as we saw in Section IV.6, he takes the orbits of the inner planets to be smaller than the orbit of the sun, so that the centers of their epicycles are definitely not the same.

[15]The reader should remember that L_S, the mean longitude of the sun, is also the longitude of the epicycle center in Figure II.1.

the elongation by giving its absolute value and then saying whether the planet was east or west of the sun, and I shall follow his practice.[16]

We will need to have some idea of the accuracy with which a Greek astronomer could find the time when a planet had a maximum elongation. We start with his accuracy in measuring longitude, as shown by the errors in Hipparchus's star catalogue. After we eliminate the bias in longitude that appears in Ptolemy's version of the catalogue, we find that the standard deviation of a measured longitude [*CCP*, p. 238] is about 22′. However, the inner planets are probably brighter than many of the stars in the catalogue, and it is plausible that the accuracy was better for brighter objects. As a talking point, let us use 15′ as the standard deviation in measuring longitude.

For Mercury, I have calculated that its elongation changes from its maximum value by 15′ in about 3 days, so it is reasonable to take this as the standard deviation in the time of the maximum, if we find the time by direct observation. For Venus, the time required to change by 15′ is somewhat more than 4 days, say 4½ days, and I shall use this as the standard deviation for Venus.

However, if we are trying to find the time when some measured quantity has its maximum value, we should not do so by measuring it near the maximum, because it changes slowly there. Instead, we should find the times when it takes on equal values on opposite sides of the maximum, and average these times. That is, we should use the averaging method described in Section II.9. Without specifying a particular procedure, we cannot estimate the accuracy obtained this way, but it is much better than we would get by direct measurement.

Ptolemy was certainly acquainted with the averaging method for finding a maximum, because he uses it in circumstances that I shall take up in Section VII.2, and it is likely that earlier astronomers also knew it. However, it does not seem likely that they used it to find the maximum elongations that Ptolemy cites. For example, take an observation of Mercury that Ptolemy quotes in his Chapter IX.7. After giving the year and day, Ptolemy says: ". . . Mercury was three moon's to the north of the brightest star in the tail of Capricorn.[17] . . . that time was. . . at dawn." Ptolemy goes on to calculate the elongation at this time and states that this was a maximum elongation.

This does not sound like a determination of the maximum by the averaging method, although it is possible that the elongation was measured on a set of days spanning the maximum, that the day of the maximum was found by the averaging method after the maximum had

[16]Instead of saying that the planet is east or west of the sun, he sometimes says whether it was observed at sunrise or sunset.

[17]δ Capricorni.

occurred, and that the measurement quoted was the one made on the day found this way. However, when we come to examine the errors in the measured times, we shall find that they are several days, so it seems unlikely that the averaging method was used. It is more likely that the observers simply estimated when the maximum elongation occurred, with no attempt to use a more sophisticated technique.

We can now turn to what Ptolemy means by a planetary apogee. When he refers to the apogee of the sun or the moon, he means a position in which the distance of the body from the earth is a relative maximum, but he does not mean this when he refers to the apogee of a planet. Instead, he means the position of the center of the epicycle when it is farthest from the earth.

In the simple epicycle model shown in Figure II.1, the value of the maximum elongation is the same whether the planet is east or west of the sun, and it is the same regardless of the value that the mean longitude L_S has at the time of the maximum. That is, the maximum elongation is independent of the position of P at the time of the maximum. However, in Ptolemy's planetary models, the deferent curve is not a circle centered on E, and its radius is a definite function of L_S.

This means that the value of the maximum elongation depends upon the value of L_S at the time the maximum occurs. Further, for a given value of L_S, the value of the elongation when it is a maximum depends upon whether the planet is east or west of the sun. Still further, these statements are actual observational facts that do not depend upon the properties of some planetary model. I think the first statement is obvious while the second is not, and I have not thought of a simple explanation of it. I shall ask the reader to accept it, or to consult the detailed calculations in *CCP* [Section X.3 for Mercury and Section XI.3 for Venus] that establish it.[18]

Thus the maximum elongation itself has maxima and minima. In order not to become mired in hopelessly long circumlocutions, let me adopt the symbol E for a value of the maximum elongation when the planet is east of the sun and W for the maximum elongation when it is west. Then E and W themselves are functions of the mean longitude L that exhibit their own maxima and minima.

When we refer the maxima and minima of E and W to the position of apogee, an interesting fact emerges that is approximately the same for both Mercury and Venus. The positions where the maxima and minima of E and W occur are approximately the same for both planets, and the approximate positions are given in Table VI.2. The maxima are sym-

[18]It would be fairly easy to see this point if the elongation were referred to the true position of the sun. The difficulty in understanding arises from the fact that Ptolemy refers the elongation to the position of the mean sun (Equation VI.5).

TABLE VI.2

POSITIONS OF THE MAXIMA AND MINIMA OF *E* AND *W*

Phenomenon	Position Referred to Apogee, *degrees*
Maximum *E*	− 120
Maximum *W*	+ 120
Minimum *E*	+ 60
Minimum *W*	− 60

metrically located with respect to apogee and lie about 120° from it on either side. The minima are also symmetrically disposed with respect to apogee and lie about 60° from it.

This is an important point. As we shall see in the next chapter, Ptolemy devises a complicated model for Mercury but he has a simpler model for Venus. Many discussions of the subject say that Ptolemy was forced to his complicated model for Mercury because the positions of the maxima and minima given in Table VI.2 are not 0° and 180°, and that the tabulated values arise because of the large eccentricity of Mercury's (heliocentric) orbit. Actually the phenomenon is about the same for both Mercury, with an eccentricity of about 0.206, and for Venus, with an eccentricity of about 0.007. The explanation for the phenomenon must lie elsewhere, and the values in Table VI.2 are not the reason why Ptolemy adopted a complicated model for Mercury. I shall discuss his reason at length in Section VII.6.

4. The Order of Development of the Theories

Pedersen [1974, p. 311] writes that we ". . . do not know the order in which Ptolemy developed the various planetary theories in the Almagest." It seems to me that this is not quite correct. The theories of all the other planets depend upon a result that Ptolemy establishes with Mercury, and the theories of all the outer planets depend further upon a result that he establishes with Venus. Thus we may say that in at least some major respects Ptolemy developed his planetary theories in the order that he presents them. This order is: Mercury, Venus, Mars, Jupiter, and Saturn.

Ptolemy's theories of the outer planets are basically identical, differing only in the numerical values of the various parameters. Thus, from the conceptual viewpoint the theories of the outer planets amount to only one theory, and the question of order is meaningless. Ptolemy could have finished his treatment of the outer planets in any order. However, it seems likely to me that he would finish his study of one planet before taking up another, because it is easier that way. I see no reason to doubt that he developed his theories in the order of presentation in the *Syntaxis*.

Pedersen [1974, p. 307] and other writers have tried to study the order in which Ptolemy finished his planetary theories by studying the dates of the latest observations that he used for each planet. It is interesting that the dates of the latest observations tend to follow the inverse order of presentation in the *Syntaxis*. Specifically, the latest dates of the observations used for each planet are: 141 February 2 for Mercury, 140 July 30 for Venus, 139 May 30 for Mars, 139 July 11 for Jupiter, and 138 December 22 for Saturn. Pedersen concludes from these dates that it is ". . . more likely than not that Ptolemy dealt first with the superior planets, and afterwards with the inferior ones."

These writers, who wrote before the appearance of *CCP*, naturally assumed that Ptolemy's "observations" were genuine. Now we know that they were fabricated, and any argument based upon the dates must fail. Since the "observations" were fabricated, the order of the dates involved has nothing to do with the order in which the "observations" were made up. It is even possible that the *Syntaxis* was finished before the date of the latest "observation" given in it, and certainly any particular planetary theory could have been completed before the date of the latest "observation" used to establish it.

Ptolemy, Hipparchus, and other Greek astronomers worked from a long background of Babylonian astronomy, but their astronomy differed from the Babylonian in at least two important ways. First, the Babylonian astronomers developed rather elaborate computational schemes, but, so far as we know, they never based their schemes upon physical models as the Greeks did. Second, the main Babylonian interest in the planets was to calculate the times and positions of various phenomena such as passage through opposition or the turning points,[19] and they had relatively little interest in calculating the positions of the planets at some arbitrary instant in time. In contrast, the main goal of the Greek astronomers was to be able to calculate the position of a planet as a continuous function of time.

Pedersen [1974, p. 329] says that the adoption of this goal was ". . . an intellectual feat of the highest importance." The adoption of

[19] I shall define these terms in Section IX.2.

this goal has certainly had intellectual consequences of the highest importance, but I am not sure that we should describe the event itself as a high intellectual feat. It was probably the Greeks, not the Babylonians, who invented judicial astrology, and it was the astrologers who first needed to calculate the position of a planet at any arbitrary time. It seems likely to me that this need of Greek astrology, rather than high intellectual considerations, is what led to this new goal of Greek astronomy.

CHAPTER VII

THE PARAMETERS OF MERCURY

1. Ptolemy's Model for Mercury

Ptolemy's theory of the moon has twelve free parameters, as we saw in Section III.2. However, two of these are the distance and the apparent diameter, while three others are connected with the motion in latitude. Thus only seven parameters are needed to specify the moon's motion in longitude. Ptolemy's model of Mercury, on the other hand, needs eight parameters to specify its motion in longitude. His model for Mercury is the most complicated model in the *Syntaxis*.

Pedersen [1974, p. 261] writes: "There is no fundamental unity in Ptolemy's theories of the five planets. A common geometrical model would have been preferable, but Mercury in particular refused to conform to the same scheme as the other planets—as we know now because of the exceptionally great eccentricity of its orbit." Later [p. 311] he writes: ". . . it is a plausible hypothesis that he set out to construct the theory of Mercury along very much the same lines as that of Venus" but ". . . he very soon ran into unexpected difficulties which could be resolved only by a model which is . . . as complicated . . . as that of the lunar theory."

I believe that the first of these quotations gives a slightly incorrect impression of the situation. Actually Ptolemy uses only two models for all five planets known to him. One model serves for Venus and the outer planets, while a second model is used for Mercury.[1] However, as I have shown [*CCP*, Chapter X], the need for a separate model for Mercury does not arise from the "great eccentricity" of its (heliocentric) orbit. Instead it arises from theoretical errors that Ptolemy made in trying to use

[1]That is, Ptolemy uses only two models to account for the geocentric motion of five planets. Copernicus uses four models to do the far simpler job of explaining the heliocentric motion of six planets.

some observations of Mercury, and I shall develop further support for this conclusion in the last section of this chapter. Ptolemy would have gotten far more accurate results for Mercury by using a model that is even simpler than the model he used for the other planets.[2] I do agree with the second quotation, but with the proviso that the main source of difficulty is not Mercury's eccentricity; it is Ptolemy's understanding of the situation.

Now we can turn to Ptolemy's model of Mercury, which is shown in Figure VII.1. The circle shown is the epicycle, with its center at *P*. The point ☿ is the position of Mercury, and the angle γ is the anomaly of Mercury, if we give "anomaly" the meaning that Ptolemy did. The anomaly increases in the direction shown in the figure. In the figure, I have drawn the epicycle with a fairly large radius, to indicate that its radius r is not to be considered as a small parameter.

Ptolemy takes the radius r of the epicycle to be a constant, so the complexity of his model concerns the location of the center P of the epicycle. To locate it, we start with point E, which represents the earth.[3] The line E♈ is the direction of the vernal equinox and the line from E to A is in the direction that Ptolemy calls apogee. Angle a is the longitude of apogee.

Line ZP rotates uniformly about Z with the angular velocity n, the mean motion of the sun, in the direction indicated by the arrow. The angle α has no particular name, but it is directly related to the mean longitude L_S of the sun by the condition

$$\alpha = L_S - a. \tag{VII.1}$$

However, there is no angle in the figure that is equal to L_S. The line DF also rotates uniformly about D in such a way that the angle ADF is always equal to α. During the motion, the distance PF stays constant. Thus we have a model that is reminiscent of the lunar model in some ways.

I think it is clear from the figure that the distance from E to P has its greatest value when P lies along the line EA, so that EA is the direction in which the deferent has its greatest radius. However, the smallest radius does not occur in the opposite direction, although I think this property of the model is not obvious. Instead, the smallest radius occurs when α is about 120° and 240°. Ptolemy expresses this fact by saying that Mercury

[2]In Section VII.5, I shall calculate the accuracy that results from using the same model for Mercury that was used for the other planets.

[3]Ptolemy determines the parallax of the sun, as we saw in Section II.8. In contradiction, he neglects the parallax of the planets, even though Mercury and Venus are closer than the sun, and Mercury is sometimes as close as the moon, in his scheme of things. As a result, he treats the earth as a point in his planetary theories.

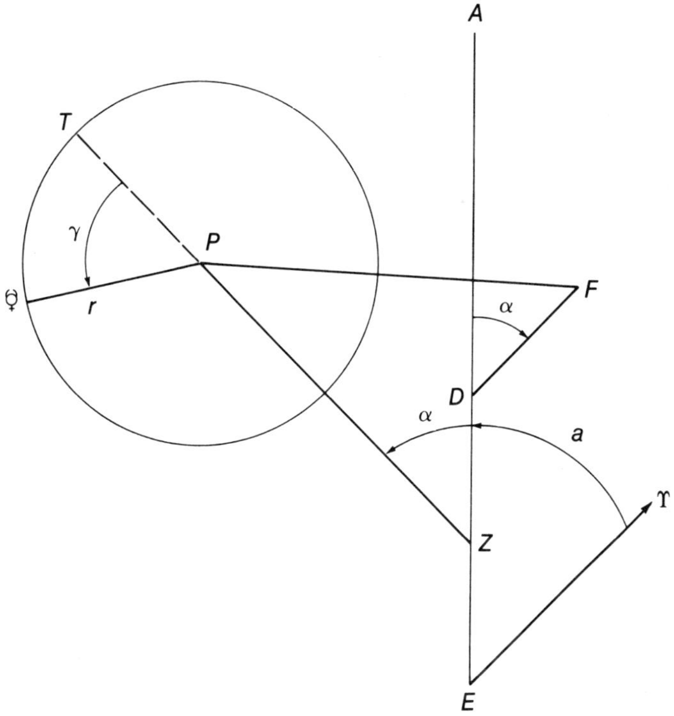

Fig. VII.1—Ptolemy's model for the motion of Mercury. *E*♈ is the direction of the vernal equinox. The planet ☿ moves uniformly around the circle whose center is *P*, the mean position of the planet. *P* rotates at a uniform rate around *Z*, but its motion is not circular. A line *DF* rotates around *D* in such a way that the two angles marked α always remain equal. The distance *PF* remains constant during the motion. Point *E* is the earth.

is "perigeed" twice in each revolution. He means, of course, that the center of the epicycle goes through two minima in each of its revolutions.

Instead of being the point where the radius of the deferent is a minimum, the direction opposite to *EA* is actually the direction in which the radius has a local maximum.

I shall use the term "first eccentric" to denote the distance *EZ*, and I shall let it equal e_1. I shall call *ZD* the second eccentric, e_2, and I shall call *DF* the crank radius, e_3. Finally, I shall denote the constant distance

PF by ρ, without giving it a more particular name. I think it will be useful to summarize this terminology and these symbols:

$$\begin{aligned} e_1 &= EZ = \text{first eccentric,} \\ e_2 &= ZD = \text{second eccentric,} \\ e_3 &= DF = \text{crank radius,} \\ \rho &= PF = \text{constant distance.} \end{aligned} \tag{VII.2}$$

Ptolemy considers the possibility that the longitude a of apogee may change at a constant rate, which I shall denote by a'. We can now enumerate all the parameters in Ptolemy's model for the longitude of Mercury:

γ_0, the anomaly at the reference epoch;
γ', the motion in anomaly;
a_0, the longitude of apogee at the reference epoch;
a', the rotation rate of apogee;
e_1, the first eccentric;
e_2, the second eccentric;
e_3, the crank radius; and
r, the epicycle radius.

Some distance in the model must be chosen arbitrarily, since Ptolemy does not try to consider planetary distances in the *Syntaxis*. Ptolemy uses two choices for the arbitrary distance at different stages in his determination of the parameters. Initially, he takes the sum $\rho + e_1 + e_2 + e_3$ to be 120. Later he takes the constant distance ρ to be 60.

Further parameters are needed to specify the motion of the planet in latitude. Ptolemy's treatment of planetary latitudes is not very satisfactory, and a study of his treatment does not contribute much to the purposes of this work. Therefore I shall ignore the planetary latitudes.

2. The Longitude of the Apogee of Mercury

In Chapter IX.7 of the *Syntaxis*, Ptolemy first uses two measurements of E and two of W, which he claims to have made himself, to find the position of the apogee of Mercury in his own time. He then uses three measurements of E and three of W made about four centuries earlier to find the position of apogee at the earlier time. From these results, he finds the values of a_0 and a'.

Table VII.1 summarizes these observations. The first column gives an identifying number that I shall use in discussing the observations. The

TABLE VII.1

OBSERVATIONS USED TO ESTABLISH THE APOGEE OF MERCURY

Identifying Number	Date	Mean Longitude of Sun, *degrees*	Elongation,[a] *degrees*
1	−261 Feb 12	318;10	25;50 W
2	−261 Apr 25	29;30	24;10 E
3	−256 May 28	62;50	26;30 E
4	−261 Aug 23	147;50	21;40 E
5	−236 Oct 30	215;10	21;00 W
6	−244 Nov 19	234;50	22;30 W
7	132 Feb 2	309;45	21;15 E
8	134 Jun 4	70;00	21;15 W
9	138 Jun 4	70;30	26;30 E
10	141 Feb 2	310;00	26;30 W

[a]From the mean sun rather than the true sun.

meanings of the other columns should be obvious. If the reader should want the measured longitude of Mercury, he can find it from the table using Equation VI.5.

I also listed these observations in Table X.1 of *CCP*.[4] There I listed them in the order in which Ptolemy presents them in the *Syntaxis*. However, it is obvious that Ptolemy did not use them in the order of presentation, and that he actually used the old observations before he used his own. In Table VII.1, I list the observations in the order in which Ptolemy obviously used them. Thus the identifying number used here for a particular observation is not the same as the number I used in *CCP*.

Ptolemy starts with observation 1, in which the elongation was 25;50 degrees west, made at a time when the mean longitude L_S of the sun was 318;10 degrees. If he could find an observation made when the elongation was 25;50 degrees east, the line *EA* in Figure VII.1 would bisect the two values of L_S. However, he says, he cannot find an old observation with this elongation, so he uses two that bracket it. These are observations 2 and 3. He interpolates between them and finds that *E*, the maximum elongation east, would be 25;50 degrees when L_S is 53½ degrees. The average of this and 318;10 degrees is 185 5/6 degrees, so apogee is either at this value or at 5 5/6 degrees.

[4]In *CCP* I also tabulated the name of the observer and the error made in measuring the longitude of Mercury.

Ptolemy then confirms this result, and settles which value belongs to apogee and which belongs to the opposite direction, by means of observations 4, 5, and 6. He interpolates between observations 5 and 6 to find that L_S would be 224;10 degrees when W equals 21;40 degrees. The average of this and the value of L_S from observation 4 is exactly 186°. It is also clear, from comparing the sizes of the elongations in the first six observations, that apogee is 186° rather than 6°. This confirms his first result within 10 minutes of arc, even though the time of a maximum elongation (Section VI.3 above) is uncertain by about 3 days and the value of L_S is correspondingly uncertain by about 3°.

However, matters are even worse than this. The correct value of apogee at the time of these observations was about 214°, so Ptolemy has made two determinations that agree within 10′ while being in error by 28°. The explanation comes from Table VI.2. Since apogee was at about 214°, the value of L_S when E was maximum was about 94°. Hence there were two values of L_S at which E took on the desired value of 25;50 degrees to match the first observation in Table VII.1, and these are symmetrically disposed about a value of L_S somewhere near 94°. The first observation is symmetrical to one of these positions but not to the other. The position that is symmetrical in fact had L_S equal to about 110°, not to the value of 53½ degrees that Ptolemy found by interpolation between observations 2 and 3. Since he used a value of L_S that was too small by almost 60°, Ptolemy made an error of almost 30° in finding apogee. A reasonable error would have been about 3° or less.

In other words, Ptolemy knew that he could find a maximum of a function by the method that I called the averaging method in Sections II.9 and VI.3 above, but he did not understand it well enough to apply it correctly. This, then, is the origin of Ptolemy's value of apogee: He intended to derive it from valid observations made about four centuries before his time, but he made a serious theoretical error in doing so.

After he made this error, Ptolemy confirmed the value of apogee by observations 4, 5, and 6. It is not plausible that a set of genuine observations would have confirmed his first erroneous result so accurately, so there must be some fabrication connected with this set of observations. My suspicion is that the times given for all three observations are genuine, and that Ptolemy chose this set because the corresponding values of L_S gave him the value of a (186°) that he wanted. I suspect that the measured longitude of Mercury, and hence the elongation calculated from it, is genuine for observations 5 and 6, and that Ptolemy simply altered the longitude and elongation of Mercury in observation 4 to the values he wanted. However, there is no rigorous way to test this suspicion, and the reader should treat the entire set of observations as if they were fabricated.

After finding that a equaled 186° at a time four centuries before his own time, Ptolemy then uses observations 7 through 10 to make two independent determinations of a in his own time, and he finds that it is almost exactly 190°. Thus a' equals 1 degree per century.

By using observations 7 through 10, Ptolemy simultaneously confirms the enormous error he has already made in the value of a and "proves" that a increases at a rate that exactly equals the precession of the equinoxes. It is clear that these observations are fabricated. It is also clear that Ptolemy chose the value of a' arbitrarily, to give it the round value of 1 degree per century and also to make it exactly equal to the precession. As a result, the orbit of Mercury remains fixed with respect to the fixed stars in his system.

I referred to this matter in Section II.5 above. There we saw that he chose to keep the apogee of the sun fixed with respect to the equinoxes, so that it precesses with respect to the stars and to the planetary orbits.[5] His reason was that it would have taken a moderately tedious computation to give a for the sun a value different from what Hipparchus had found. For Mercury, on the other hand, we see that it was a simple matter for him to put apogee anywhere he wanted to. He chose to keep apogee fixed with respect to the stars, and he doubtless thought that doing so made for an elegant picture of the heavens.

3. The Eccentrics, the Crank Radius, and the Epicycle Radius

Ptolemy finds these four parameters by using the four observations of E and W that are listed in Table VII.2. According to him, Theon made the observation that is numbered 13 and he himself made the others. He first uses observations 11 and 12 to find the epicycle radius r, the sum $\rho + e_3$, and the sum $e_1 + e_2$.[6] In doing so, he adopts the convention that $\rho + e_1 + e_2 + e_3$ equals 120. The meanings of these symbols are explained in relations VII.2 above. Ptolemy does this in Chapter IX.8 of the *Syntaxis*, and I have analyzed his work in Section X.3 of *CCP*.

In observation 11, Ptolemy claims that Mercury was at maximum elongation west on the morning of 134 October 3, when L_S was 189;15 degrees. He found the longitude of Mercury to be 170 1/5 degrees, so that the value of W was 19;03 degrees. In observation 12, he claims that

[5]Ptolemy adopts $a' = 1$ degree per century for all the planets, as we shall see in the appropriate places.

[6]I start with observation number 11 in Table VII.2 in order to avoid confusion with the ten observations in Table VII.1.

TABLE VII.2

OBSERVATIONS USED TO FIND THE ECCENTRICS, THE CRANK RADIUS, AND THE EPICYCLE RADIUS OF MERCURY

Identifying Number	Date	Mean Longitude of Sun, *degrees*	Elongation,[a] *degrees*
11	134 Oct 3	189;15	19;03 W
12	135 Apr 5	11;05	23;15 E
13	130 Jul 4	100;05	26;15 E
14	139 Jul 5	100;20	20;15 W

[a] From the mean sun rather than the true sun.

Mercury was at maximum elongation east at sunset on 135 April 5, when L_S was 11;05 degrees. He found the longitude of Mercury to be 34;20 degrees, so that the value of E was 23;15 degrees.

Note the suspicious accuracy with which he gives the longitude of Mercury in observation 11, where he uses a fraction of a degree that he almost never uses. Note that one of these observations was made when L_S was close to Ptolemy's value of the apogee a and the other was made when L_S was almost diametrically opposite.

When P, the center of the epicycle, is near apogee ($L_S = 189;15$ degrees), the distance from E to A is $\rho + e_1 + e_2 + e_3$. When P is in the other direction, the distance is $\rho - e_1 - e_2 + e_3$. Therefore, if we remember that the sine of the elongation is the epicycle radius r divided by the distance, and if we remember that we take the first sum above to equal 120 by convention, the pair of observations gives us:

$$\begin{aligned} r &= 39;09, \\ \rho + e_3 &= 109;35, \\ e_1 + e_2 &= 10;25. \end{aligned} \qquad \text{(VII.3)}$$

In Chapter IX.9 of the *Syntaxis*, Ptolemy turns to a measurement of E and one of W made when L_S had the same value and was 90° from apogee (observations 13 and 14). The fact that he cannot find observations meeting these conditions exactly means that the following discussion is not exact and requires slight corrections. Since we are interested in the ideas involved more than in the exact values in this work, I shall assume in this discussion that L_S was exactly 100° (90° from 190°) for both observations.

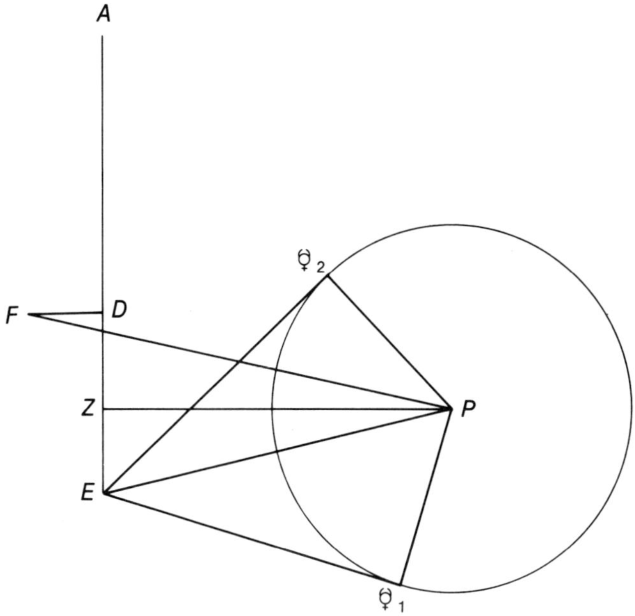

Fig. VII.2—Ptolemy's model of Mercury for certain specific observations. Ptolemy claims that he has two observations made when line *ZP* was almost exactly perpendicular to line *EA*. Mercury was at the point $☿_1$ on one occasion and at $☿_2$ on the other.

The configuration of Ptolemy's model for these two observations is shown in Figure VII.2. The point marked $☿_1$ in the figure is the position of Mercury when it had its maximum west elongation, which was 20;15 degrees; Ptolemy claims that he made this observation (number 14) himself on 139 July 5. The point marked $☿_2$ is the position of Mercury when its maximum east elongation was 26;15 degrees. Ptolemy says that this observation (number 13) was made by Theon on 130 July 4.

Since the line of sight was tangent to the epicycle on both occasions, the triangles that share the side *EP* are right triangles. In order to analyze the figure, we need two results from elementary geometry: The angle $☿_1 E ☿_2$ subtended by the epicycle equals 46;30 degrees, the sum of the two elongations, and the angle *ZPE* is half their difference, namely 3;00 degrees. With the aid of these relations, it is easy to find the distances *ZE* and *DF*, which are e_1 and e_3, respectively. Ptolemy finds $e_1 = 5;12$ and $e_3 = 5;13$. These results, when used in Equations VII.3, give $e_2 = 5;13$ and $\rho = 104;21,49$.

Ptolemy points out that all three *e*'s are essentially equal, so he takes them to be 5;12, and he rounds ρ to 104;22. Now, however, Ptolemy decides to drop the convention that ρ plus the sum of all the *e*'s is to be 120, and instead he adopts the convention that $\rho = 60$. This requires him to rescale *r* and all the *e*'s. *r* now becomes 22;30,26 and each of the *e*'s becomes 2;59,22. Ptolemy rounds these to give him his finally adopted values:

$$\rho = 60, \quad r = 22;30, \quad e_1 = e_2 = e_3 = 3;00 . \quad \text{(VII.4)}$$

The *e*'s are obviously round values and *r* is also obviously round to someone who is familiar with the sexagesimal system: $r/60 = 3/8$ exactly.

Now we turn to the origin of these parameters. I think it is clear that Ptolemy intended from the beginning to have all three *e*'s come out equal. Thus we can take $e_1 = e_2 = e_3 = e$ from the beginning, and thus we are really involved with only two parameters, not four, in this section. These parameters are *e* and *r*.

On pp. 292ff of *CCP*, I investigated several ways in which Ptolemy might have been led to his values of *r* and *e*, without reaching any conclusions. I mentioned that he could have found them by analyzing observations 1 and 2, which may well be genuine, and I commented that this explanation was perhaps the most likely of all the ones I suggested. However, I did not carry out a test of that possibility.

It is relatively simple to deduce *r* and *e* from a pair of observations. Imagine that the three distances *EZ*, *ZD*, and *DF* in Figure VII.1 are all equal to *e*. Imagine further that the line is drawn in from *E* to ☿, and let β denote the angle *AE*☿. This angle is the longitude of Mercury minus *a*, while α is the mean longitude of the sun minus *a*. Thus the elongation equals β minus α. An observation furnishes values of β and α once *a* is known. Let *R* denote the distance *PZ*.

It is easy to show that the *x* and *y* coordinates of point *F* are:

$$x = e \sin \alpha, \quad y = 2e + e \cos \alpha.$$

The *x* and *y* coordinates of point *P* are:

$$x = -R \sin \alpha, \quad y = e + R \cos \alpha.$$

The square of the distance from *F* to *P* is then

$$(e \sin \alpha + R \sin \alpha)^2 + (e + e \cos \alpha - R \cos \alpha)^2 .$$

In his "second normalization" Ptolemy takes this distance to be 60, but it is easier to write the relations if we take it to be 1. A little manipulation yields the equation

$$R^2 - 2eR(\cos\alpha + \cos 2\alpha) + 2e^2(1 + \cos\alpha) = 1. \quad \text{(VII.5)}$$

If we neglect quantities of order e^2, the solution of this is

$$R = 1 + e(\cos\alpha + \cos 2\alpha). \quad \text{(VII.6)}$$

The mathematics available to Ptolemy would have allowed him to find this result.[7]

With point P now located accurately, we can easily write down the x and y coordinates of ☿, and apply the condition that $\tan\beta = -x/y$ for this point. After some elementary manipulation, this leads to:

$$\begin{aligned} e[\sin\beta + (\cos\alpha + \cos 2\alpha)\sin D] + r &= -\sin D, \quad D<0, \\ e[\sin\beta + (\cos\alpha + \cos 2\alpha)\sin D] - r &= -\sin D, \quad D>0. \end{aligned} \quad \text{(VII.7)}$$

D, we remember, is the elongation $\beta - \alpha$.

At the epoch of observations 1 and 2 in Table VII.1, apogee was at 186° according to Ptolemy's solution for this parameter. Thus, $\alpha = 132\ 1/6$ degrees and $\beta = 106\ 1/3$ degrees for observation 1. For observation 2, $\alpha = 203\ 1/2$ degrees and $\beta = 227\ 2/3$ degrees. Use of these values in Equations VII.7 gives $e = 0.057\ 347$ and $r = 0.361\ 481$. If we follow Ptolemy and take the constant distance to be 60 rather than 1, we multiply these values by 60, and at the same time we should convert to sexagesimal notation. This gives

$$e = 3;26, \qquad r = 21;41\ . \quad \text{(VII.8)}$$

There is no difficulty in assuming that Ptolemy would round the value of e to 3;00, which is the value in Equations VII.4. In the usual manner of speaking, one would not round 21;41 to 22;30. However, if a person were particularly aware of the fact that 3/8 times 60 equals 22;30, he could easily decide to change r from the odd value 21;41 to 22;30. In assessing this possibility, the reader should remember that modern students are drilled extensively in the decimal equivalents of simple frac-

[7] I do not mean that Ptolemy could have gotten this algebraic result. I mean that, given the values of e and α, he could have gone through the arithmetic operations which parallel the algebraic operations used in deriving Equation VII.6.

tions, such as the equivalence of 0.375 to 3/8 or the equivalence of 0.833 333 . . . to 5/6, and students in Ptolemy's time were probably drilled in the sexagesimal equivalents of simple fractions such as 3/8.

Whatever the reader may think of the possibility that Ptolemy rounded 21;41 to 22;30, there is no conceptual problem in the origin of his values of r and e. He first set all the individual e's equal to the same value e, although this is a bad choice, as we shall see in Section VII.5. He then found r and e by analyzing two observations that had been made by other astronomers, and doubtless rounded the results after he had done so.

4. The Anomaly and the Motion in Anomaly

Ptolemy has now found all the parameters in his model except γ_0 and γ'; he has not needed them yet because so far he has worked entirely with values of maximum elongation. To find these parameters, he first finds the longitude of Mercury on 139 May 17 by observing a conjunction of it with the moon, and also by measuring its separation from Regulus (α Leonis). He combines this with an observation of Mercury made by the Alexandrian astronomer Dionysios on −264 November 15 to find the values of γ_0 and γ'.

I have already discussed this determination in Section VI.2, with a few main conclusions: (a) Ptolemy fabricated the observation he claims to have made on 139 May 17, (b) the value he adopts for γ' cannot be derived from the observations, and (c) the value of γ' does not agree with the period relations given in Table VI.1, although Ptolemy claims to have derived the period relations from the observations quoted.

In any case, there is no difficulty in seeing how Ptolemy got his values of γ_0 and γ'. He simply adopted a value of γ', perhaps from Babylonian sources; this is the value in Table VI.1. He then found from Dionysios's observation that the anomaly was 212;34 degrees at 6 A.M. on −264 November 15; this is 176 312.75 days after the era of Nabonassar. Multiplying the number of days by the value of γ' from Table VI.1 tells us that the change in anomaly was 1 521 complete revolutions plus 190;39 degrees. Hence the anomaly at the era of Nabonassar was 21;55. In other words, Ptolemy got γ_0 by combining an accurate value of γ' and an observation, which is probably genuine, that had been made about four centuries before his own time.

5. The Accuracy of Ptolemy's Model for Mercury

Studying the accuracy of Ptolemy's models is not one of the purposes of this work; I have already done that in *CCP*. However, the Mercury situation is so extraordinary that I think the reader will enjoy a brief reprise of the situation. Further, I have obtained some new results on the matter.

In Section X.4 of *CCP* I made an extensive comparison of Ptolemy's theory with Newcomb's,[8] which we may take as exact for the purposes of this comparison. The greatest error I found in Ptolemy's model was 7°.84, and the standard deviation of the error was 2°.99, which we may round to 3°.

I tried the effect of changing apogee to the correct value while keeping Ptolemy's other parameters unchanged. This dropped the standard deviation only to 2°.54, which is not a large improvement. I forgot to record the greatest error. In other words, as long as we keep Ptolemy's complicated crank model, with the three e's equal to each other, we cannot get an accurate representation of Mercury's motion.

I then let all the parameters vary in Ptolemy's model and found the values that give the best fit to Newcomb's theory. Apogee of course changed to about the correct value. The other parameters barely changed except for the e's. e_1 rose from 3;00 in Ptolemy's second normalization ($\rho = 60$) to about 5;14. e_2 went to $-0;92$ (note the sign), and e_3 went to 0;72. The greatest error dropped from 7°.84 to 3°.26 and the standard deviation dropped from 3° to 1°.70.

The small sizes of e_2 and e_3 suggested the experiment of setting them equal to zero, without bothering to redetermine the other parameters. When I did so, the standard deviation rose only a small amount, to 1°.73. If the reader will now look back to Figure VII.1, he will see that making $e_2 = e_3 = 0$ puts point F permanently at Z. The deferent becomes a circle, with the earth being eccentric to it by 5;14 parts out of 60. In fact, the deferent becomes the simple eccentric model of Figure II.2.

In other words, if Ptolemy had used a simple eccentric for the deferent, without the complicated crank mechanism of Figure VII.1, he would have almost halved his errors. As I said before, the great eccentricity of Mercury's heliocentric orbit is not the reason why he adopted a complicated mechanism. The reason is his theoretical error in trying to analyze the data.

There is one more interesting point concerning the accuracy. As I said in Section VII.1, Ptolemy uses one model for Mercury and a second model for all the other planets. It is interesting to see what would have happened if he had used the same model for Mercury that he did for the other planets. This model is pictured in Figure VIII.1 and described in Section VIII.1. I would rather not take the space here to explain the

[8]Found in *Newcomb* [1895a].

model, and I ask the reader to let me simply present the results, deferring their interpretation to Section VIII.1.

I have found the parameters in this model, called the equant model, that gives the best fit to Newcomb's theory of Mercury. The results are shown in Table VII.3. Two features of the results are interesting: (a) The standard deviation of the error is still 1°.67, so the use of the more complicated model of Figure VIII.1 has not helped much,[9] and (b) the first eccentric is still considerably larger than the second.

TABLE VII.3

THE PARAMETERS IN THE EQUANT MODEL THAT BEST FIT THE ORBIT OF MERCURY

Parameter	Value
Apogee, *degrees*	219;58
First eccentric (e_1)[a]	4;06
Second eccentric (e_2)[a]	1;07
Epicycle radius[a]	22;39
Anomaly at the epoch,[b] *degrees*	187;15,10
Motion in anomaly, *degrees per day*	3;06,22,57
Standard deviation of the errors, *degrees*	1.67

[a]Referred to a deferent radius of 60.
[b]At the epoch 137 July 20 rather than at the era of Nabonassar.

6. The Origin of Ptolemy's Model for Mercury

It is fully as important to ask after the origin of Ptolemy's model as it is to ask after the origin of the parameters he used in it. So we should ask: What considerations led Ptolemy to use the complicated model shown in Figure VII.1 when simpler models give much more accurate results?

Ptolemy gives an explicit answer near the end of his Chapter IX.8, where he refers again to observations 7 and 10 in Table VII.1. Both measurements were made when L_S was 310°, to the accuracy we need here. The value of E (the maximum east elongation) for this value of L_S

[9]At least it does not make matters worse, as Ptolemy's model does.

is 21;15 degrees while the value of W for the same L_S is 26;30 degrees. I pointed out in connection with Figure VII.2 that the sum of the two elongations is the angle subtended by the epicycle when its center has the stated value of L_S. Thus the epicycle, according to Ptolemy, subtends an angle of 47;45 degrees when $L_S = 310°$. It also subtends this angle when $L_S = 70°$, by symmetry, still according to Ptolemy.

On the other hand, when L_S is close to 10°, as it was for observation 12 in Table VII.2, the measured value of E was 23;15 degrees. For this value of L_S, the center of the epicycle is on the symmetry axis, and the values of E and W are equal. Thus the epicycle subtends an angle of 46;30 degrees when L_S is directly opposite the direction of apogee.

To recapitulate, when the center of the epicycle is at the end of the symmetry axis that is nearest the earth ($L_S = 10°$), the epicycle subtends an angle of 46;30 degrees, but when the center is 60° away in either direction it subtends the larger angle of 47;45 degrees. Thus it is closer to the earth at these two points than when it is on the symmetry axis. Ptolemy expresses this condition by saying that Mercury is "perigeed" twice in each revolution. More specifically, the center of the epicycle passes through two positions of perigee during each revolution (a sidereal year). These two perigees lie about 120° from apogee, or about 60° from the opposite direction.

This conclusion no more corresponds to reality than does Ptolemy's earlier conclusion that apogee is at 190° rather than 220°. If we calculate the sum of the east and west elongations as a function of L_S, using the orbital parameters that were correct in Ptolemy's time [*CCP*, Figure X.10], we find that the single maximum of about 48° comes when L_S is about 40° and that the minimum of about 42° comes when L_S is about 220°. The former should have been Ptolemy's value of perigee and the latter his value of apogee, and he should have had only a single perigee.

Further, his argument is specious since all the observations used in it were fabricated by himself. Still, it seems psychologically unlikely to me that Ptolemy would have started out with a model as complicated as Figure VII.1, which is designed with the production of two perigees as its main goal. It seems to me that only some compelling reason would have led him to undertake all this work.

In Section VII.3, I investigated the possibility that Ptolemy found the parameters r and e in his model by using observations 1 and 2 in Table VII.1. Now let us suppose instead that he started out with the equant model of Figure VIII.1, but with the special condition that the two eccentrics are equal,[10] and that he analyzed observations 1 and 2 by means of this model.

[10] As we shall see, Ptolemy "proves" that the two eccentrics are equal for this model, in parallel with what he did for his models of the moon and of Mercury.

The analysis is similar to the earlier analysis of these observations, and it requires again that we neglect terms proportional to e^2. Even so, it is simpler than the earlier analysis, and I think that the reader should be able to carry it out with the aid of the discussion in Section VIII.1 below. The important relations, which resemble Equations VII.7, are:

$$\begin{aligned} e(\sin\beta + \sin\alpha\cos D) + r &= -\sin D, \quad D<0, \\ e(\sin\beta + \sin\alpha\cos D) - r &= -\sin D, \quad D>0. \end{aligned} \qquad \text{(VII.9)}$$

I think the reader can easily derive these by making use of the first of Equations VIII.1 in Section VIII.1, with e_2 being replaced by e. β is the angle AEB in Figure VIII.1, so that the longitude of B is $\beta + a$. By iteration we can readily find a solution that is accurate through terms of order e^2, but I think it would be rather tedious to go beyond this.

I shall not give the details of the solution and shall proceed immediately to the results:

$$r = 21;14, \qquad e = 3;01. \qquad \text{(VII.10)}$$

In deriving these values, we must remember that Ptolemy thought that a equaled 186° at the epoch of the observations.

Ptolemy would probably round these values to 21;15 and 3;00, respectively, so let us do the same. His next step would be to calculate the value of the maximum elongation at the end of the symmetry axis opposite to apogee; this is the place where α and β are both 180°. The distance from E to the center of the epicycle would be $60 - e = 57;00$, and we are taking the radius of the epicycle to be 21;15. The ratio of these is the sine of the maximum elongation, which has the same value whether it is east or west when the center is on the symmetry axis. This gives

$$E = W = 21;24,54 \text{ degrees}, \qquad \text{(VII.11)}$$

which Ptolemy would probably round to 21;25 degrees. Thus he would find that the angle subtended by the epicycle was 42;50 degrees when $L_S = 6°$.

On the other hand, in observation 3 in Table VII.1, E equals 26;30 degrees when L_S equals 62;50 degrees. In order to find the angle subtended by the epicycle at this position, we need a measurement of W, but Ptolemy does not give us one. However, if it is anything bigger than 16;20 degrees (and the least value it can have anywhere is about 19°), the angle subtended at this position will be greater than 42;50 degrees, which is the value Ptolemy finds at the short end of the symmetry axis. In other words, this end cannot be perigee, as it has to be in the equant model. Instead, there is a perigee position near the point where $L_S = 62;50$ degrees, and there must be another perigee at the symmetric point. Thus Ptolemy is forced to a model that will produce this double perigee.[11]

On the other hand, suppose that Ptolemy had not made his error about the position of apogee, and that he had put it at the correct position, which is about 215° at the epoch of the observations in question. This would subtract 29° from all the values of α and β that are listed in the text between Equations VII.7 and VII.8. When we repeat the analysis using the equant model, we find

$$r = 24;20,12, \qquad e = 0;58,36. \qquad \text{(VII.12)}$$

I feel rather sure that Ptolemy would have rounded e to exactly 1. I am not sure what he would have done about the value of r. He might have rounded it to 24;20, or he might have rounded it all the way down to 24 exactly. The rest of the argument will be more conservative if we assume that he did the latter. Then, if we use $r = 24$ and $e = 1$, we get

$$E = W = 24;00 \text{ degrees} \qquad \text{(VII.13)}$$

in place of the value in Equation VII.11, and we get 48;00 degrees for the angle subtended by the epicycle. We still do not know what angle Ptolemy would have used where $L_S = 62;50$ degrees, but it is not likely that it would have exceeded 48;00. The correct value [*CCP*, Figure X.10] was about 47;46 degrees. Thus, if Ptolemy's values had been reasonably accurate, he would have found that perigee did indeed lie on the symmetry axis. That is, he would have reached the correct conclusion that there is only one apogee and one perigee, and that they are 180° apart. He would also have used the same model for Mercury that he did for the other planets.

Of course, all this argument is somewhat speculative, but it accomplishes one very important thing. It says that Ptolemy's errors about the position of apogee and about the double perigee are not independent errors, but that the double perigee followed naturally from the first error. Further, the argument is based upon data which we know were available to Ptolemy, and all the actions of Ptolemy that it calls for are ones that seem natural to him.

[11]If this analysis is correct, what Ptolemy did here stands in strong contrast to what he did with his lunar model. There, when he found that his model gave the wrong variation of the lunar distance, he kept the model and fabricated the data to fit. Here, if I am right, when he found that his first choice of the model apparently gave the wrong position of perigee, he changed his model to agree with the data. I believe this is the only instance in which Ptolemy corrected an "error" that he made. The irony is that the only time Ptolemy thought he was correcting an error, he was in fact making a new one.

There is a further interesting point about the observations. Ptolemy could have used any pair chosen from observations 1, 2, and 3, and I have done the required analysis for all three combinations. Equations VII.12 and VII.13 give the results when we use observations 1 and 2. When we use observations 2 and 3, we get $e = 2;13$, $r = 24;03$, and $E = W = 24;35,34$ degrees. Since this is bigger than 24;00 degrees, it strengthens the argument. On the other hand, if we use observations 1 and 3, we get the impossible result that $e = -1;0,28$. What would Ptolemy have done if he had happened to use this pair? I suppose that he would have concluded that the equant model does not work for Mercury, but he would not necessarily have been led to the double perigee.

In sum, Ptolemy made a blunder when he attempted to apply the averaging method for finding the position of apogee. In consequence the position that he found was in error by almost 30°. When he tried to confirm his first result by using additional data, he found that the two values did not agree. Instead of trying to find out which result was correct, he simply altered the data used in the second determination to make it agree with the first.

His error in the position of apogee then led him to the conclusion that the position of perigee was not opposite to that of apogee. Instead he concluded that Mercury has two perigee positions that are symmetrically located with respect to apogee. This conclusion in turn led him to the complicated model shown in Figure VII.1. He would have gotten much more accurate results if he had used the same model for Mercury that he used for the other planets.

Thus, the complexity of his Mercury model does not result from the large eccentricity of Mercury's orbit. It results from the initial error of analysis that Ptolemy made when he tried to find the position of apogee. The large eccentricity is responsible only for the size of the errors that are present in any model of Mercury that has a circular epicycle.

CHAPTER VIII

THE PARAMETERS OF THE OTHER PLANETS

1. The Model Used for the Other Planets

The model that Ptolemy uses for Venus and the outer planets has come to be called the equant model in the recent literature about Ptolemaic astronomy. This model is shown in Figure VIII.1. In Figure VIII.1, the deferent (which is not drawn in) is a circle whose center is at *Z*. The earth is not at *Z* but is at the point *E*. If point *P* moved uniformly around point *Z*, its motion would follow the simple eccentric model of Figure II.2. However, the motion is uniform around point *D* rather than *Z*. That is, the angle *PD*♈ increases at a uniform rate; this angle is the mean longitude of the planet. When the point *P* is at point *A*, its distance from *E* is a maximum, so *A* is the point that Ptolemy calls the apogee of the planet.

Point *B* represents the position of the planet, and *B* moves uniformly around a circle centered on *P*. The anomaly γ, which increases in the direction shown, is measured from the line *PT*, which is the extension of the line *DP*. The longitude of apogee is the angle a, while α equals L, the mean longitude,[1] minus a.

The radius *PZ* of the deferent circle can be chosen arbitrarily. In most discussions I shall follow Ptolemy and take *PZ* to be 60. In a few discussions, however, it will be more convenient to use 1. The distance *EZ* will be called the first eccentric, denoted by e_1, and the distance *ZD* will be called the second eccentric, denoted by e_2. The epicycle radius will be denoted by r as before.

This model has only two small distances (e_1 and e_2), just as the lunar model of Figure III.1 does, and this is one less than the Mercury model

[1]This mean longitude equals L_S, the mean longitude of the sun, for Venus but not for the outer planets. Hence I shall use L without a subscript in speaking of the mean longitude in connection with the equant model.

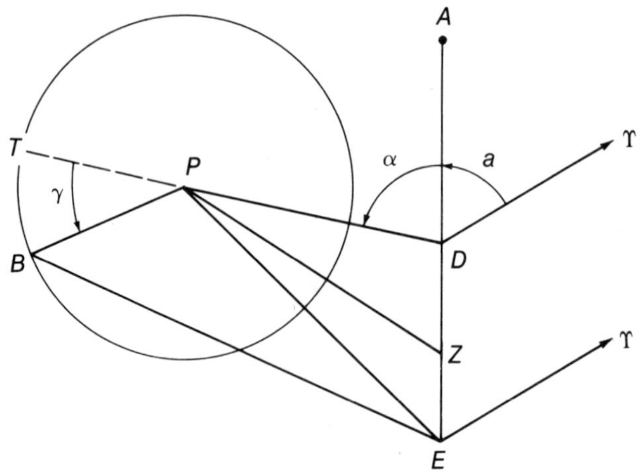

Fig. VIII.1—Ptolemy's model for the motion of Venus and the outer planets. The earth is at E and $E\Upsilon$ is the direction to the vernal equinox. The planet B rotates uniformly around P, the mean position. P rotates at a uniform rate as measured at D, while its distance from a different point Z remains constant.

of Figure VII.1. Therefore it has seven parameters,[2] which are shown in the following list:

- γ_0, the anomaly at the reference epoch;
- γ', the motion in anomaly;
- a_0, the longitude of apogee at the reference epoch;
- a', the rotation rate of apogee;
- e_1, the first eccentric;
- e_2, the second eccentric; and
- r, the epicycle radius.

I have discussed the advantages of introducing the second eccentric in considerable detail in Section XI.1 of *CCP*, and I shall only summarize them briefly here. Suppose that we have only the first eccentric e_1. If we choose it to give the correct maximum equation of the center, the maximum error in longitude is $2e_1^2$. Further, the distance varies from $1 + 2e_1$ to $1 - 2e_1$, but the variation on an elliptic orbit is only from

[2] I continue to ignore the latitudes of the planets.

$1 + e_1$ to $1 - e_1$.[3] By introducing the second eccentric, we can simultaneously reduce the errors in longitude and in the distance variation.

There are two important cases. In the first, if $e_1/e_2 = 5/3$, the equation of the center is exact through terms of order $e_1{}^2$, but there is some error in the distance variation. If $e_1/e_2 = 1$, we eliminate the error in the distance variation (to the order of $e_1{}^2$), but there is some error in the equation of the center.

The angle subtended by the epicycle at *E* is an important property of a planetary model, and this in turn is related to the distance *PE*. Further, the longitude of *P* (the angle *PE*♈ in Figure VIII.1) is also obviously important. Thus any real case would be a mixture of the two special cases, with the ratio e_1/e_2 lying between 1 and 5/3, if it were accurate to take the epicycle to be a circle. Since the correct epicycle should not be a circle, this fact distorts the limits on e_1/e_2, but the optimum ratios come close to lying within the indicated range except for Mercury. I have given the optimum ratios for Venus and the outer planets in Table XI.8 of *CCP*. As we saw in Table VII.3, the optimum ratio for Mercury is about 3.67, which is far outside the indicated range. This is presumably an effect of the large eccentricity of Mercury.

Returning to Figure VIII.1, let angle *EPZ* be denoted by δ_1, let angle *ZPD* be denoted by δ_2, and let angle *PZD* be denoted by θ_2. The relations we need to calculate the position of point *P* are given by Equations XI.1 of *CCP*, which I repeat here for the convenience of the reader:

$$\begin{aligned} \sin \delta_2 &= e_2 \sin \alpha, \\ \theta_2 &= \alpha - \delta_2, \\ \tan \delta_1 &= e_1 \sin \theta_2/(1 + e_1 \cos \theta_2), \\ \lambda_P &= L - \delta_1 - \delta_2. \end{aligned} \qquad \text{(VIII.1)}$$

In the last of these, λ_P denotes the true longitude of point *P* while *L* denotes its mean longitude. I am still using the convention that the distance *PZ* is 1 rather than 60. The first of these is the relation used in deriving Equations VII.9 in Section VII.6.

2. The Parameters for Venus

Ptolemy finds the parameters for Venus in almost exactly the same way that he found those for Mercury. The main difference concerns a', the rate of change of apogee. Ptolemy "proved" by means of observa-

[3]In this immediate discussion I take the deferent radius to be 1 and I take angles to be in radians.

TABLE VIII.1

EIGHT OBSERVATIONS OF VENUS AT MAXIMUM ELONGATION

Identifying Number	Date	Mean Longitude of Sun, *degrees*	Elongation,[a] *degrees*
1	132 Mar 8	344;15	47;15 E
2	140 Jul 30	125;45	47;15 W
3	127 Oct 12	197;52	47;32 W
4	136 Dec 25	272;04	47;32 E
5	129 May 20	55;24	44;48 W
6	136 Nov 18	235;30	47;20 E
7	134 Feb 18	325;30	43;35 W
8	140 Feb 18	325;30	48;20 E

[a] Defined as the actual longitude of Venus minus the mean longitude of the sun.

tions four centuries apart that a' is 1 degree per century for Mercury. He simply assumes this value without comment for all the other planets. Thus only six parameters remain to be found in the list given in the preceding section.

The fact that he does not find a' means that he does not need the counterpart of the six old observations of Mercury listed in Table VII.1. Instead he uses only eight observations of Venus at maximum elongation, all made in his own time.[4] These observations are tabulated in Table VIII.1, which has the same format as Tables VII.1 and VII.2. Ptolemy says that Theon made the observations numbered 1, 3, and 5 and that he made the rest himself.

The average of the mean longitudes of the sun is exactly 235° for observations 1 and 2, and it is 234;58 degrees for observations 3 and 4. Ptolemy therefore concludes that apogee is either at 55° or 235°, but he cannot tell yet which it is.

I find it suspicious that in both cases the elongations he claims to have measured himself (numbers 2 and 4) match those measured by Theon to the minute of arc, even when the number of minutes is the unusual value of 32.

As I showed in *CCP* [p. 311], the error in Ptolemy's value of a is about 4°. *Gingerich* [1980] describes the result as "astonishingly good",

[4] In fact, he says near the beginning of his Chapter X.1 that he cannot find any old observations from which he can determine apogee. Since he had found at least six old observations of Mercury that were suitable, and since Venus is far easier to observe than Mercury, this is rather unlikely.

and he refers to the "very sensitive geometry" involved. Actually the result does not appear particularly good to me. As we see from the calculations in Section VI.3, the error is just about the size we expect from a single pair of mean elongations that were found by the method of direct observation. The use of two pairs of elongations, giving two independent determinations, should have made the error smaller by a factor of $\sqrt{2}$. Further, if the elongations had been determined by the averaging method of Sections II.9 and VI.3,[5] the error should have been still smaller, perhaps of the order of 1° or less. An accuracy of 4° is in fact rather poor.

Actually, I do not believe that Ptolemy made the observations in Table VIII.1 that he claims to have made, although the observations attributed to Theon may be genuine. Aside from the suspicious agreement of the elongations just mentioned is the suspicious agreement of the two determinations. They agree within 2 minutes of arc under circumstances where we expect and find an error of 4°. The probability that two independent determinations would both be in error by a standard deviation while agreeing within less than 0.01 standard deviation is about 0.002. Therefore at least one observation from each of the first two pairs in the table has been fabricated. The natural inference is that the observations attributed to Theon may be genuine while those which Ptolemy claims to have measured are fabricated.

With Ptolemy's way of finding apogee, his *a priori* chance of success is 1 in 2. This is exactly what happened. He failed the first time he tried, with Mercury, but he succeeded the second time.

He now turns to observations 5 and 6, which were made with the center of the epicycle close to 55° and 235°, respectively. We see from the size of the elongations that 55° is apogee while 235° is perigee. At apogee, the distance from E to P is $60 + e_1$ and at perigee it is $60 - e_1$.[6] Thus we have

$$r = (60 + e_1) \sin 44;48,$$
$$r = (60 - e_1) \sin 47;20.$$

Ptolemy's solution of these relations, which is reasonably accurate, is

$$e_1 = 1\ 1/4, \qquad r = 43\ 1/6. \qquad \text{(VIII.2)}$$

[5]The times used for the maximum elongations, for both Mercury and Venus, were all either sunrises or sunsets. This is unlikely if they had been found by the averaging method. Thus the times as well as the size of the error both indicate that the maxima were found by the direct method, which is certainly inferior to the averaging method.

[6]I am now returning to the convention of taking *PZ* to be 60.

He finds the second eccentric e_2 by using observations 7 and 8 in Table VIII.1. These are observations of maximum east and west elongations made with exactly the same value of L, something that he did not accomplish with Mercury. The value of L is not exactly 90° from apogee, but this is a minor matter.

The configuration of the model of Figure VIII.1 at the times of these observations is shown in Figure VIII.2. This is somewhat simpler than its Mercury counterpart (Figure VII.2) because there is one less parameter. Either of the equal angles $♀_1 EP$ and $♀_2 EP$ equals half the sum of the elongations, namely 45;57,30 degrees, and the angle DPE is half the difference, or 2;22,30. With these angles known, it is easy to find that the distance DE, which equals $e_1 + e_2$, is almost exactly 2½, so that

$$e_1 = e_2 = 1\tfrac{1}{4}. \qquad \text{(VIII.3)}$$

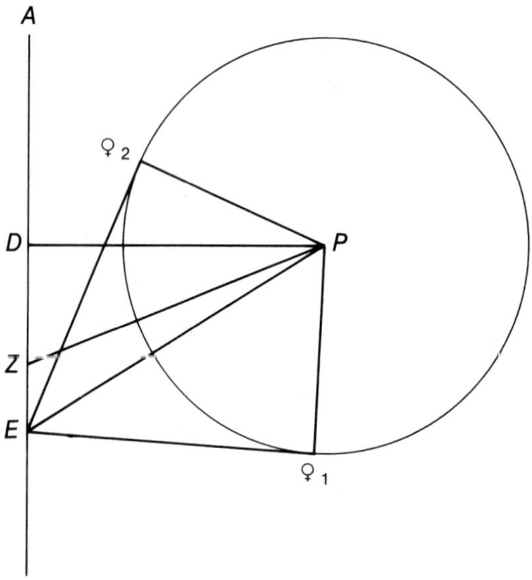

Fig. VIII.2—Ptolemy's model of Venus for certain specific observations. Ptolemy claims that he has two observations made when line DP was almost exactly perpendicular to line EA. Venus was at the point $♀_1$ on one occasion and at $♀_2$ on the other.

Again Ptolemy has made the small distances come out equal, as he did with the moon and with Mercury. Actually they should not be equal, and both should be considerably smaller than 1¼. The best values [*CCP*, p. 311] are

$$e_1 = 0;46,23, \quad e_2 = 0;52,58. \qquad \text{(VIII.4)}$$

The best value of r is

$$r = 43;22,06, \qquad \text{(VIII.5)}$$

so Ptolemy's r is fairly accurate.

Finally he finds γ_0 and γ' by using two observations. He claims that he found the longitude of Venus at 4¾ hours after midnight on 138 December 16 both by observing a conjunction of it with the moon and by measuring its separation from Spica (α Virginis).[7] He also used an observation of a conjunction of Venus and η Virginis that Timocharis made half an hour before sunrise on −271 October 12.

The same remarks I made in Section VII.4 apply here and to all the planets: (a) The observation that Ptolemy claims to have made was fabricated, (b) the value he adopts for γ' cannot be derived from the observations, and (c) the value of γ' does not agree with the period relations in Table VI.1, although Ptolemy claims that he derived those relations from the observations in question.

Then, after adopting a value of γ' from some source that is now lost, Ptolemy finds γ_0 from the observation of Timocharis. He finds that the anomaly was 252;07 at the time of the observation, which was 173 721¾ days after the era of Nabonassar. He finds that the change in anomaly in this time, aside from the complete circles, is "very nearly 181°." (I find 180;59 degrees.) Hence the anomaly at the era of Nabonassar was 71;07 degrees.

3. General Remarks About the Outer Planets; Mean Opposition

Ptolemy continues to use the model of Figure VIII.1 for the outer planets, but there are nonetheless some differences between them and Venus. One of them is that the angle *PD*♈ in the figure is no longer equal to the mean longitude L_S of the sun. It is now the mean longitude L of

[7] This is really two independent observations made at the same time.

the planet, and for an outer planet the mean longitude is the same whether we use a geocentric or a heliocentric picture.[8] Instead, as I indicated in Section VI.1, the direction from P to B is the mean longitude of the sun, so we have

$$L_S = a + \alpha + \gamma = L + \gamma. \tag{VIII.6}$$

Ptolemy knew and used this relation.

Ptolemy continues to use the elongation of a planet, which I shall denote by D, and the definition of the elongation continues to be that of Equation VI.5:[9]

$$D = \lambda - L_S. \tag{VIII.7}$$

D is not limited in value as it is with an inner planet, but can increase without limit.

When D equals 180°, we have the condition that Ptolemy calls oppositon to the mean sun, or mean opposition. That is, the true position of the planet is in opposition to the mean position of the sun rather than to its true position. Presumably one finds the time of a mean opposition by measuring the longitude λ on a succession of nights, calculating the mean longitude L_S of the sun at the time of each measurement, and calculating the elongation D from Equation VIII.7. One continues the process until D crosses 180° and then interpolates to find the time when it was exactly 180°.

The importance of a mean opposition for a planetary theory is shown by Figure VIII.3. This shows the configuration of the equant model at the time of a mean opposition. The important thing is that the anomaly γ is such that the planet B lies along the line from the center P of the epicycle to the earth E. It is not obvious that this is so. This is a theorem that Ptolemy proves in Chapter X.6 of the *Syntaxis*, and I have also given a proof of it [*Newton*, 1976, Section XIII.6]. This theorem means that at a mean opposition we can actually measure the longitude of the epicycle's center even though there is no body there to send light to our eyes. I shall not repeat the theorem here, but I will say that the reader should not be surprised by it; the mean longitude L_S of the sun is connected with the value of the anomaly γ through Equation VIII.6.

The longitude λ_P of the center of the epicycle is related to the parameters of the orbit by Equations VIII.1. Since λ_P is measured at the time of a mean opposition (in fact $\lambda_P = \lambda$ there), one can use these

[8] I do not mean to imply that Ptolemy pointed out this fact.
[9] λ is the geocentric longitude of the planet.

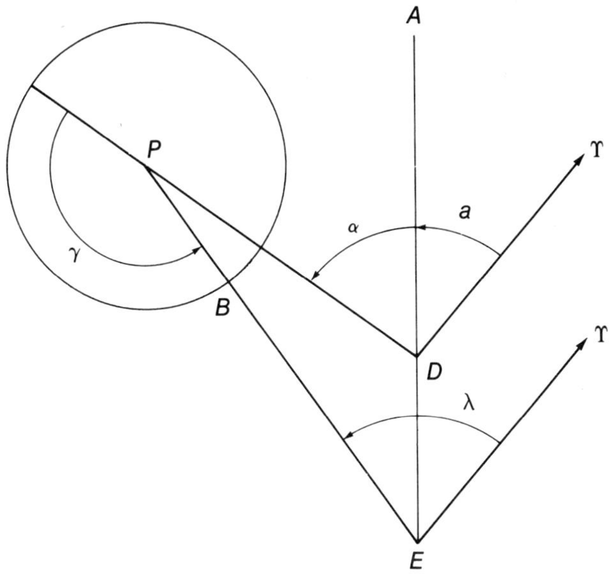

Fig. VIII.3—Mean opposition of an outer planet. At a mean opposition, the direction from the earth *E* to the mean sun is directly opposite to the direction *EP*. It is a theorem to prove that, in Ptolemy's model, the true position *B* of the planet lies along the line *EP*, so that the mean planet and the true planet are simultaneously in opposition to the mean sun.

measurements to find the relevant parameters. I shall return to this point in the next section.

We saw earlier that Ptolemy "proves" that a', the rate of change of apogee, is exactly 1 degree per century for Mercury, and that he takes this value for granted in his Venus theory. He does the same for the outer planets, without further comment.

He also "proves" that the two eccentrics in the equant model are equal for Venus. At the beginning of Chapter X.6 of the *Syntaxis*, he claims that he has shown the same equality for each of the outer planets by means of observation. However, he gives no details, and he takes $e_1 = e_2 = e$ from the very beginning of his treatment of the outer planets.

Ptolemy treats each outer planet in an identical fashion, so we need to treat only one of them in order to follow what he does for all of them. I shall choose Jupiter as an example.

4 The Parameters for Jupiter

At the beginning of Chapter XI.1 of the *Syntaxis*, Ptolemy gives three mean oppositions of Jupiter. The first came at 1 hour before midnight on 133 May 17, when the longitude of Jupiter was 233;11 degrees. The second came at 2 hours before midnight on 136 August 31, when the longitude of Jupiter was 337;54 degrees, and the third came at 5 hours after midnight on 137 October 8, when the longitude was 14;23 degrees. Since the oppositions were presumably determined by interpolation, it is not suspicious that the longitudes are given to a precision of a minute of arc.

From these three observations, Ptolemy finds a, the position of apogee in his own time,[10] the eccentric e (he takes the two eccentrics in the model of Figure VIII.1 to be equal, as I have already noted), and the value of the anomaly at the time of the third opposition. In doing this, he has to calculate the change in anomaly that occurs between the observations. Strictly speaking, he cannot do this because he has not yet found the value of γ', which actually constitutes a fourth unknown. However, over the short span of the data, he does not need an accurate value, and almost any preliminary value is good enough. Thus it is legitimate to take γ' as known for the moment.

Ptolemy's solution for the three parameters takes up 12 pages in the cited translation of the *Syntaxis*, and I have made no attempt to follow it except to note that it is a method of successive approximations, starting from the approximation that $e = 0$. He goes through the process twice for both Jupiter and Saturn, for both of which e is about 3 parts out of 60. For Mars, whose e is about twice as large, he goes through the process three times.[11] As I noted in Section III.4, this problem is a distant relative of the problem that Ptolemy has already solved for the sun and moon, and that Hipparchus had solved before him. For the reader who is interested in the method, *Pedersen* [1974, pp. 172-176] gives an explanation of the solution for the moon, but not of the more difficult solution for the outer planets.

Earlier [*Newton*, 1976, pp. 442-445] I posed the problem in essentially algebraic rather than geometric terms and outlined the solution, commenting as I did so that it was quite difficult. There I was interested mainly in duplicating Ptolemy's results, and it did not occur to me to ask if there were a simpler way to solve the problem. In the course of writing the present work, it did occur to me to ask this question, and almost immediately I realized that there is a simple and elegant method of solution. This is the method that I called the method of the seasons in Sections II.5

[10] He neglects the amount of precession that occurs during the 4-year span of the data.
[11] This solution takes up about 16 pages.

and III.4. Since this method requires the use of longitudes that differ by multiples of 90°, we cannot apply it to the data that Ptolemy presents. However, I think it is important to make a test of the method for the outer planets, using synthetic data in order to do so. I shall present this test in Appendix B rather than here, in order not to interrupt the study of Ptolemy's analysis.

Since Ptolemy does not choose his observations of the outer planets in a way that lets him use the elegant method of the seasons, we must accept this fact and continue with his analysis. Next [*Syntaxis*, Chapter XI.2] he uses two observations that he made at the same time, namely 5 hours after midnight on 139 July 11. At this time, he observed that Jupiter was directly north of the center of the moon. From this, after correcting the lunar position for the equation of time and for parallax, he finds that the longitude of Jupiter was 75;45 degrees. At the same time, he measured the separation of Jupiter from α Tauri, and from this measurement he also finds that the longitude of Jupiter was 75;45 degrees.

I do not see how there can be any doubt that this observation was fabricated. The moon and the stars both shared the bias in longitude of about 1°.1, as I have already discussed in Chapter V, but there are still the other sources of error to consider. The error of observation was about 20′ for both measurements involved, yet Ptolemy finds agreement exact to the minute. He does this in spite of the fact that he applied the equation of time with the wrong sign [*CCP*, p. 263]. Further, since the elongation of the moon was about 315° at the time of the observation, his value of the parallax was considerably too large. It is not credible that his errors of observation would have accidentally cancelled these two theoretical errors, in addition to cancelling the large periodic errors (standard deviation about 35′) in his theory of the moon, and that they would have done so in the analogous observations of all the other planets.

From this fabricated observation, Ptolemy concludes that the radius r of the epicycle of Jupiter is 11;30 parts out of 60.

Ptolemy has now found four of the five parameters in the model for Jupiter, and the only remaining parameter is γ'. In Chapter XI.3 of the *Syntaxis*, he pretends to find this from an old observation of Jupiter. He says that Jupiter occulted δ Cancri in the morning of −240 September 4, but he does not give the observer, the place, or the hour.[12] In his calcula-

[12]The corresponding observation of Saturn was made on −228 March 1. *Pedersen* [1974, p. 272] says that this observation was made "by an unknown Babylonian astronomer". However, in his use of this observation, Ptolemy clearly assumes that it was made in Alexandria. I believe that Pedersen was misled by the fact that Ptolemy gives the year of the observation as the "82nd year according to the Chaldeans". I do not know what Ptolemy meant by this "era of the Chaldeans", which would have to lie in the year −309 or −310 if he did not make an error in the number 82. The nearest era I can find is the era of Seleucus, which was −311 September 1. Seleucus was the one of Alexander's generals who became the ruler of Babylonia after the death of Alexander.

tions he takes the time of the observation to be sunrise in Alexandria. He deduces that the anomaly of Jupiter was 77;02 degrees at the time of the observation, while it had been 182;47 degrees at the time of the mean opposition of 137 October 8. Thus, he says, paying due regard to the number of complete revolutions of anomaly in the interval (345), the value of γ' is 0;54,9,2,46,26,0 degrees per day. As I pointed out in Section VI.2, this is not the value of γ' that one gets from the data, nor is it the value that one derives from the period relations in Table VI.1. The analogous statement is true for all the planets, both inner and outer.

5. The Origin of the Parameters

I get the distinct impression that Ptolemy was getting tired of the job by the time he came to Venus and the outer planets. His discussions of the problems involved for the sun, the moon, the precession, and Mercury are extensive, and he makes multiple determinations of many of the parameters. With Venus, on the other hand, he uses six less observations than he did for Mercury, and the only old observations he uses are the ones needed to get γ'. Altogether he quotes 12 observations in the process of getting five parameters.[13] However, he uses the twelfth observation only to show whether a certain quantity was increasing or decreasing at the time of the eleventh observation, and two others were two independent measurements of the longitude of Venus, so only ten observations are used numerically in the process of finding five parameters. a is the only parameter that he determines more than once.

Matters are even worse for the outer planets. He uses six observations to find four parameters. However, two of the observations were made at the same time, to give him a longitude of the planet when it is not in opposition, so he uses only five positions of the planet to find four parameters. Further, he quotes only one old observation.

Let us finish up the discussion of the outer planets first, since it will be simpler than the discussion of Venus, and let us continue to use Jupiter as an example. It is easy to see where Ptolemy got his value of γ_0, the anomaly at the era of Nabonassar. From the observation made on −240 September 4 he found that the anomaly was 77;02 degrees. This observation was made 185 006¾ days after the era of Nabonassar, so the change in anomaly, using Ptolemy's value of γ' from Table VI.1, was 463 full revolutions plus 290;58 degrees. Hence the value of γ_0 is 146;04 degrees.

[13] His model contains seven parameters, but he adopts two of them (a' and γ') arbitrarily, and he actually finds only five parameters from the data quoted.

Since Ptolemy uses six observations to find four parameters, we can establish the fabrication of only two observations, and we cannot test the other four. We have already noted that Ptolemy finds the position of Jupiter on 139 July 11 in two independent ways. The exact agreement, in the face of observational error and of two theoretical errors Ptolemy made in reducing the data, shows that one of the observations was fabricated, but we cannot tell which one it is.[14]

Let us turn again to the observation of −240 September 4, when the anomaly was 77;02 degrees, and from it let us calculate the anomaly at the time of the third opposition of Jupiter, which is 5 hours after midnight on 137 October 8. The time interval between the observations is 137 732 days plus 23 hours. If we multiply this by the value of γ', we find 345 full revolutions plus 105;46 degrees for the change in anomaly, and the anomaly at the time of the third opposition is 182;48 degrees. By his analysis of the three oppositions, however, Ptolemy finds that the anomaly was 182;47 degrees at this time. If we take 15′ as the standard deviation in measuring the longitude of a planet, the uncertainty in a value of the anomaly must be measured in degrees, not minutes of arc, because of the small size of the epicycle radius. Agreement within 1′ is not plausible.

Hence either the observation of −240 September 4 or the one of 137 October 8 is fabricated, and rigorously we can prove nothing more than this. Since Ptolemy has been known to fabricate observations attributed to other astronomers, we must even admit the possibility that the earlier observation was fabricated. However, if either observation is genuine, it is plausible to take the genuine one to be that of −240 September 4.

The observation of −240 September 4 is the only observation of Jupiter which Ptolemy uses that he does not claim to have made himself and, if it is genuine, it is the origin of Ptolemy's value of γ_0. There is no way to determine the origins of the other three parameters that Ptolemy actually finds for Jupiter on the basis of data in the *Syntaxis*, because there are no such data.

However, we should remember that we already know the origins of the parameters a' and the ratio e_2/e_1 of the two eccentrics in the equant model. Ptolemy claims to have proved that a' equals 1 degree per century for Mercury, but we may be sure that he simply adopted this value because it is a round number. He also claims to have proved from observations that the ratio of the eccentrics is 1 for Venus and for all the outer planets, but we may be sure that he adopted the value 1 because it is a round number. Thus we know the origins of three of the seven

[14]In earlier discussions, I said that the agreement showed that *these* observations were fabricated, and I have no doubt that both were. As a matter of logic, however, we can prove only that one of them was.

parameters that are actually present in the model for Jupiter, but we can find no more.

This does not mean that we should accept three of Ptolemy's observations as genuine, however. We have shown that two of the five observations he claims to have made were fabricated,[15] and it seems unlikely to me that he suddenly changed his entire *modus operandi* and genuinely made the other three observations.

Turning now to Venus, we may be sure that Ptolemy intended to make the ratio e_2/e_1 equal to unity from the beginning, so there is only one eccentric value to be found; I shall call it e. We also know that he adopted $a' = 1$ degree per century arbitrarily. He has taken γ' from some unknown source, but we know that the Babylonians had accurate values for many mean motions, so there is no mystery in the fact that Ptolemy's value of γ' is rather accurate. I have already described how he found the value of γ_0 from an observation made on -271 October 12. This leaves only the origins of the parameters a_0, e, and r to be explained.

In a comparable situation with Mercury, I found the origin of a_0 in three old observations which Ptolemy analyzed by means of a symmetry argument.[16] Then, with a_0 known, I analyzed some old observations to find e and r, with the results shown in Equations VII.10. The value of e found there is almost exactly Ptolemy's value. The value of r is not quite the same as his value, but we can think of reasons why he might have altered 21;14 to 22;30.[17] In the case of Venus, there are five observations that Ptolemy attributes to other astronomers, and we should see whether he might have obtained his values of a_0, e, and r from them. The five observations are the ones numbered 1, 3, and 5 in Table VIII.1, which are attributed to Theon, and the two observations listed in Table VIII.2, which are attributed to the early Alexandrian astronomer Timocharis. These are the ones that Ptolemy used, along with his observations made on 138 December 16, to find γ' and γ_0. (We should remember that he only pretends to find γ' but that he does actually find γ_0 from these data.) I give the observations in Table VIII.2 the identifying numbers 11 and 12.

These five observations do not have the balance of elongations needed to let us find apogee by the symmetry method, and to solve for a_0, e, and r simultaneously is a difficult job. However, this is not necessary, since we are not trying to find the parameters *ab initio*. We are merely trying to find if some of the observations are consistent with

[15]In judging Ptolemy's conduct, it does not matter whether the observation of -240 September 4 was fabricated or whether the observation of 137 October 8 was. For simplicity of discourse, I shall assume that the latter was fabricated and that the former is genuine.

[16]Perhaps I should say misanalyzed.

[17]We should not forget the possibility that he might have made an error in computation.

TABLE VIII.2

TWO ANCIENT OBSERVATIONS OF VENUS

Identifying Number	Date	Mean Longitude of Sun, *degrees*	Elongation,[a] *degrees*
11	−271 Oct 12	197;03	−42;53
12	−271 Oct 16	200;59	−42;09

[a] Defined as the actual longitude of Venus minus the mean longitude of the sun. These are not maximum elongations. A minus sign means a west elongation.

Ptolemy's parameters. Thus we can assume Ptolemy's value of a_0 and solve for only e and r from the observations. In fact, if we take a_0 as known, each observation furnishes the linear relation between e and r that is given in Equations VII.9 in Section VII.6. For our five observations 1, 3, 5, 11, and 12, these relations are, respectively:[18]

$$\begin{aligned} 1.039\ 597\, e + r &= 0.734\ 322, \\ 1.403\ 247\, e + r &= 0.737\ 670, \\ -0.694\ 710\, e + r &= 0.704\ 634, \\ 1.381\ 708\, e + r &= 0.680\ 508, \\ 1.321\ 453\, e + r &= 0.671\ 074. \end{aligned} \qquad \text{(VIII.8)}$$

From these five equations, we can form ten independent solutions for e and r, which are given in Table VIII.3. Since the observations are divided into two groups, I have divided the table into three groups. The first group of observations consists of numbers 1, 3, and 5, which were made by Theon in Ptolemy's own time, and the second group consists of numbers 11 and 12, which were made by Timocharis slightly more than four centuries earlier. In the first group of solutions I have put the three solutions that we obtain from observations 1, 3, and 5; in the second group of solutions I have put the single solution that we can get from observations 11 and 12; and in the third group of solutions I have put the six solutions that we can get by combining any one of observations 1, 3, or 5 with either of observations 11 or 12.

In studying Table VIII.3 we should remember that Ptolemy's values are e = 1;15 and r = 43;10. We should also remember that the best-

[18] We must remember that we can find the longitude of Venus from the data in Table VIII.2. We must further remember that α in the equations is the mean longitude of the sun minus a while β is the actual longitude of the planet minus a. Ptolemy takes a to be 55° in his own time and 50;55 degrees at the time of the observations in Table VIII.2. D is the elongation, as defined in the table.

TABLE VIII.3

SOME SOLUTIONS FOR THE ECCENTRIC AND THE EPICYCLE RADIUS OF VENUS

Observations Used	Eccentric[a]	Epicycle Radius[a]
1,3	0;33,09	43;29,06
1,5	1;01,38	42;59,30
3,5	0;56,41	42;56,04
11,12	9;23,38	27;51,03
1,11	−9;26,17	53;52,16
1,12	−13;27,51	58;03,24
3,11	159;14,15	−179;11,21
3,12	48;51,07	−24;17,28
5,11	−0;41,50	41;47,37
5,12	−0;59,55	41;35,03
Ptolemy's solution	1;15	43;10

[a] Referred to a deferent radius of 60.

fitting values (Equations VIII.4 and VIII.5) are e_1 = 0;46,23, e_2 = 0;52,58, and r = 43;22,06.

The first group of solutions gives values of r that are close to both Ptolemy's value and to the best-fitting value. The values of e are close to the best-fitting values of e_1 and e_2 but are significantly smaller than Ptolemy's value of e. This suggests strongly that the first group of observations is genuine and, correspondingly, that Ptolemy did not use this group of observations in finding e. I do not see any plausible way in which the values of the eccentric in Table VIII.3 might have been rounded or altered into 1;15, even though 1;15 ÷ 60 = 1/48.

The single solution in the second group is quite inaccurate but is nonetheless sensible. We have no right to expect much accuracy from this solution because the two observations behind it were only four days apart. The fact that the solution is sensible in spite of the short time span suggests that the second group of observations is also genuine.

The trouble comes with the third group of solutions, in which we combine each observation from the first group in turn with each observation from the second group. No solution in this mixing of observations makes any sense at all. Since each group of observations by itself is self-consistent, their mutual incompatibility suggests that one group or the other has been fabricated. However, before we reach this conclusion, we should investigate carefully the effect of observational errors, as well as the effect of Ptolemy's error in the value of a' and other matters.

Thus we cannot find the origins of the parameters a_0, e, and r for Venus. As I have said before, this does not mean that we must accept Ptolemy's observations as genuine. It merely means that he derived the parameters from observations that are now lost, or that he simply adopted them from the work of others.

One final thing should be noted about apogee in relation to Equations VIII.1, which give the longitude of the center of the epicycle in the equant model. When a value of the time is given, it is straightforward to calculate the mean longitude L and to subtract the apogee a from it; this gives the angle α. With $e_1 = e_2 = e$ known, it is then simple to find λ_P, the longitude of the epicycle center. Then, by calculating the anomaly from the time, one can easily go from the center of the epicycle to the position of the planet. In other words, it is easy to fabricate a position of the outer planet for any value of apogee, and so it was easy for Ptolemy to give a' any value he liked. It was only for the sun that he needed to take $a' = 0$.

I pointed out in connection with the sun in Section II.5 that the greatest labor Ptolemy saved himself by taking $a' = 0$ was probably not the labor of fabricating the data. Rather, it was probably the labor of finding the parameters from the fabricated data. If a' were equal to zero for the sun, the seasons would have the same lengths every year. This means that Hipparchus's calculation of the solar parameters from the lengths of the seasons, which was made for a time near -140, would be valid without change in Ptolemy's time. Thus, if he chose a' for the sun, Ptolemy could take over Hipparchus's solution without change and save himself some labor, even while pretending to use his own observations.

He could not do this for a planet, however, because a planet would have two independent mean motions even if a' were equal to zero. This means that the process of finding the parameters from observations could not be the same for observations in Hipparchus's time and in Ptolemy's. Thus, when Ptolemy pretended to use his own observations of the planets, he was forced to make a new solution for the parameters, regardless of the value of a'. Hence he could choose any value of a' that he wanted without changing the amount of labor involved, and he chose to make a' exactly equal to 1 degree per century for all the planets.

CHAPTER IX

CONCLUSIONS AND DISCUSSION

1. A Tabulation of the Parameters

We are now in a position to tabulate all the parameters in the *Syntaxis* and to summarize what we know about their origins. Table IX.1 contains this information. Since Ptolemy treats all the outer planets in the same way, it is superfluous to consider more than one of them. Hence I continue to use Jupiter as a representative example, just as I did in the preceding chapter.

One interesting feature of Table IX.1 is the way our information becomes less certain as we move from the sun to the outer planets.[1] For the sun, we know the origin of all the parameters except the ratio of the apparent solar diameter to the minimum lunar diameter. As we saw in Section IV.6, Ptolemy chose to make the solar diameter equal to the minimum lunar diameter instead of to the average, as earlier astronomers had done. I speculated that he did this for some reason connected with his *Planetary Hypotheses*, but more research on this point is clearly needed. It may be that we have lost the information that would be needed to settle the question. I see no way at present to test the speculation.

For the moon, there are several parameters for which we do not know the specific source. However, we know that Hipparchus did a great deal of research about the moon, and the parameters that are marked "Hipparchus?" in Table IX.1 are probably either due directly to Hipparchus or were derived by Ptolemy from observations that Hipparchus had made.

The situation concerning Mercury is still fairly satisfactory, but there is much that is unknown about the parameters for Venus, and we know the origin of only three of the seven parameters for Jupiter.

[1]This is an additional reason for thinking that Ptolemy developed his theories in the order of their presentation in the *Syntaxis*. See Section VI.4.

TABLE IX.1

PTOLEMY'S PARAMETERS AND THEIR ORIGINS

Parameter	Value	Origin
A. SUN		
Tropical year, *days*	365¼ − (1/300)	Hipparchus
Mean longitude at the epoch, *degrees*	330;45	Hipparchus
Apogee, *degrees*	65;30	Hipparchus
Precession rate of apogee	0	Ease of fabrication
Eccentric	2;30	Hipparchus
Obliquity, *degrees*	23;51,20	Eratosthenes; makes the longest day 13½ hours on the tropics
Longitude of Alexandria, *degrees*	30;58	$\tan^{-1}$ 0.6
Ratio of solar to lunar distance	19.1	Aristarchus
Ratio of apparent diameter to minimum lunar diameter	1	Perhaps the need to make his model of the universe come out right
Choice of model		Standard Greek model
B. MOON		
Synodic month, *days*	29;31,50,8,20	Babylonian
Mean longitude at the epoch, *degrees*	41;22	Hipparchus?
Anomaly at the epoch, *degrees*	268;49	Hipparchus?
Mean motion in anomaly, *degrees per day*	13;3,53,56,17,51,59	Hipparchus, followed by rounding errors in fabricating data
Epicycle radius	5;15	Rounded from 60 sin 5°; 5° was itself rounded
Argument of the latitude at the epoch, *degrees*	84;15	Hipparchus?
Mean motion in argument of the latitude, *degrees per day*	13;13,45,39,48,56,37	Hipparchus, followed by rounding errors in fabricating data
First small distance in his model	10;19	Error of convenience in analyzing an observation made by Hipparchus

TABLE IX.1 (continued)

Parameter	Value	Origin
Ratio of second to first small distance	1	Round number
Inclination of orbit, *degrees*	5	Round number
Mean distance at the syzygies, *earth radii*	59	Probably Hipparchus
Apparent diameter at greatest distance, *degrees*	0;31,20	Probably an older result
Choice of model		Error in choosing to vary deferent radius instead of epicycle radius
C. THE EQUINOXES		
Precession rate, *degrees per century*	1	Round number
D. MERCURY		
Anomaly at the epoch, *degrees*	21;55	Analysis of a known observation
Mean motion in anomaly, *degrees per day*	3;6,24,6,59,35,50	Period relations
Apogee four centuries before Ptolemy's time, *degrees*	186;0	Theoretical error in analyzing known observations
Precession of apogee, *degrees per century*	1	Equal to precession of equinoxes
First eccentric	3;0	Probably rounded from the analysis of known observations
Ratio of second to first eccentric	1	Round number
Ratio of crank radius to first eccentric	1	Round number
Epicycle radius	22;30	Probably rounded from the analysis of known observations
Choice of model		Flows from Ptolemy's error of 30° in apogee

TABLE IX.1 (continued)

Parameter	Value	Origin
E. VENUS		
Anomaly at the epoch, *degrees*	71;07	Analysis of a known observation
Mean motion in anomaly, *degrees per day*	0;36,59,25,53,11,28	Period relations
Apogee in Ptolemy's time, *degrees*	55;0	Probably rounded from the analysis of unknown observations
Precession of apogee, *degrees per century*	1	Equal to precession of equinoxes
First eccentric	1;15	Probably rounded from the analysis of unknown observations
Ratio of second to first eccentric	1	Round number
Epicycle radius	43;10	Probably rounded from the analysis of unknown observations
Choice of model		Gives approximately the correct variation of the apparent size of the epicycle
F. JUPITER		
Anomaly at the epoch, *degrees*	146;04	Analysis of a known observation
Mean motion in anomaly, *degrees per day*	0;54,9,2,46,26,0	Period relations, followed by an error in arithmetic
Apogee in Ptolemy's time, *degrees*	161;0	Unknown
Precession of apogee, *degrees per century*	1	Equal to precession of equinoxes
First eccentric	2;42	Unknown
Ratio of second to first eccentric	1	Round number
Epicycle radius	11;30	Unknown
Choice of model		Gives approximately the correct variation of the apparent size of the epicycle

However, this should not cause any doubt that Ptolemy fabricated his observations of Venus and the outer planets. There must have been many observations of these planets available to him, and certainly many Babylonian observations of these planets are still available to us today. I see no reason to doubt that Ptolemy found the parameters in question by analyzing some genuine observations which he does not cite in the *Syntaxis*, or by simply appropriating earlier results.

Another interesting feature of Table IX.1 is the role that rounding plays in the values of many of the parameters. The obliquity was chosen to make the length of the longest day on the tropics come out to 13½ hours exactly,[2] and the latitude of Alexandria is $\tan^{-1} 0.6$, a value that is in error by 15′. For the moon, Ptolemy rounded the maximum equation of the center (at the syzygies) to 5° and then got the epicycle radius by calculating 60 sin 5° and rounding the result to 5¼. All the models except that of the sun have at least two eccentrics or small distances of some sort, and Ptolemy always chooses the ratio of these parameters to be exactly 1, even though this is often not a good choice.

Finally, Ptolemy chose to use the round number 1 degree per century for the precession of the equinoxes, even though the data available to him gave a value of around 1.3 or 1.4 degrees per century. He then arbitrarily chose to make the precession of apogee exactly equal to the precession of the equinoxes for all the planets.

We should not be surprised by the idea that Ptolemy would choose certain parameters because they were round numbers. A mind that could be an avid supporter of astrology would not boggle at the idea that important parameters of the cosmos were necessarily round numbers. Perhaps the surprise is that he did not force more parameters to have round values.

In at least two instances, the value that Ptolemy gave to a parameter was apparently determined by convenience. He took zero for the precession rate of the solar apogee in order to avoid carrying out two sets of computations. Also, in order to ease the task of finding the maximum equation of the center for the moon from an observation made by Hipparchus, he stated that the parallax was zero even though it actually had a fairly sizeable value. This error determined the value that Ptolemy decided upon for the maximum equation.

In summing up our knowledge of the origins of the parameters, it is useful to divide them into two classes. In one class we have the parameters, such as the mean motions or the mean anomaly at the epoch, that could have come from the analysis of real data, including those that may have been taken over from earlier astronomers. We cannot always

[2]But we should remember that the calculation was not done quite correctly. See Section II.7. We should also remember from Section V.4 that Ptolemy failed to mention two other ancient Greek values of the obliquity.

find the data involved, or earlier statements of the parameters, but it seems to me that this does not affect our understanding of what Ptolemy did. It just means that these specific sources are lost along with most of ancient astronomy.

The second class is the important one for our understanding of Ptolemy's work. It consists of those parameters, such as the precession rates of the equinoxes and of the solar and planetary apogees, that could not have come from the analysis of real data. It is valuable that we have found origins for all these parameters, with the exception of the ratio between the solar and lunar diameters.

In other words, with the exception just noted, we now understand how Ptolemy obtained all the parameters in the *Syntaxis*, even though we cannot locate the sources of his specific data in some cases.

2. Origin of the Models, Particularly of the Equant

The model itself that is used to describe the motion of a particular body is something like a parameter: Out of an infinity of possible models, or at least out of a large family, it is the task of an astronomer to choose the one that is capable of giving the best fit to the data. Thus Table IX.1 summarizes our information about the origin of the various models that Ptolemy used.

The simple eccentric model used for the sun was a standard model that had been used in Greek astronomy centuries before Ptolemy. The lunar model, as we saw in Section IV.1, was the result of a bad guess that Ptolemy made. In order to produce a model that would explain the evection, Ptolemy had to let either the radius of the epicycle or the radius of the deferent be a function of position, and he had to guess which one to try first. He chose to vary the radius of the deferent first, and he cannot be criticized for trying it. But when he found that it was the wrong guess, he did not try the other possibility. Instead he fabricated data to make it appear that his first guess was right.

The model for Mercury, as I showed in Section VII.6, was also the result of making a wrong guess, in this case when he was finding its apogee, and again he fabricated data to make it appear that his guess was right. Then, when he examined some data using his seriously wrong value of apogee, it appeared that perigee was not in the direction opposite to apogee; by symmetry, this meant that there had to be two positions of perigee. Hence he had to invent a model that gave two perigees. We should give him credit for being able to invent a model that met this requirement even though there was no basis for the requirement itself.

Turning now to the equant model used for Venus and the outer planets, astronomers and historians have always assumed that Ptolemy was the inventor of it. Part of the basis for this assumption is that no presently known source older than the *Syntaxis* contains a description of the equant. However, this is not a sufficient basis for attributing the model to Ptolemy. I gave an analogous situation in *CCP* [p. 368]. For a long time a certain important theorem in spherical trigonometry was attributed to Ptolemy simply because it did not occur in any earlier known work. However, we know now that it does occur in an earlier work by Menelaos that first became available in a European language in 1936 [*Menelaos*, ca. 100].[3]

The rest of the basis for crediting Ptolemy with the equant is that he specifically states in his Chapter IX.2 that he invented it. The early observations of the planets were poorly organized, he says. Further, they tended to be times of turning points (I will define this term in a moment) and other phenomena that are not well suited to determining the parameters of the motion. Hipparchus, he goes on to say, "left us more examples of accurate observations than he ever got from his predecessors." Hipparchus also arranged the observations in a useful way and showed that they could not be explained either by a simple epicycle, by a simple eccentric, or even by the combination of the two,[4] but he left the matter there, still according to Ptolemy. It was Ptolemy himself who realized that the eccentric model had to be modified by the introduction of what I call the second eccentric, he goes on to say.

As I explained in Section VIII.1, the equant model allows us to describe rather accurately both the motion in longitude and the variation in distance of the epicycle center. We need an accurate representation of the distance because it governs the angle subtended by the epicycle. We can measure this angle directly for an inner planet by the use of maximum elongations. The analogous quantity for an outer planet is the length of the retrograde arc, which I shall now define.

Most of the time a planet moves from west to east among the stars. As an outer planet nears the time of an opposition, however, or as an inner planet nears the time of an inferior conjunction, the planet comes to

[3] *Toomer* [1970, p. 190] writes that Ptolemy ". . . of all ancient authors is most inclined to give credit where it is due . . ." He wrote this before many recent results in research on Ptolemy were available. Now we know quite differently. Ptolemy presented Menelaos's theorem without acknowledgment, he appropriated Hipparchus's solar theory without acknowledgment, and he appropriated Hipparchus's star catalogue without acknowledgment. With regard to the latter, *Pedersen* [1974, p. 258] writes: "Our general impression of his moral and intellectual integrity would be damaged beyond repair if we had to believe that he simply derived his catalogue from a previous work by Hipparchus without the slightest acknowledgement of the fact." This, unfortunately, is what we must now believe.

[4] *Pedersen* [1974, p. 266] says that Hipparchus probably realized the impossibility of accounting for planetary motion by a simple eccentric, which implies, it seems to me, that he did not try the device of having the epicycle carried on an eccentric deferent. If this is what Pedersen means, it seems to me that it contradicts what Ptolemy explicitly says.

a stop. This time, or the position of the planet when it happens, is called the first turning point. After the turning point, the planet moves from east to west, passes through opposition (or inferior conjunction, as the case may be), and comes to a second turning point. Following this, it resumes its normal eastward motion. The arc followed by the planet between its turning points is called the retrograde arc.[5]

Because of the eccentricities of the sun's orbit and the heliocentric orbits of the planets, the lengths of the retrograde arcs vary from one opposition to another. Someone may have discovered that the equant model accounts for this variation better than an epicycle moving on a simple eccentric, and this discovery may have been the basis for preferring the equant model. At least that may be the implication of Ptolemy's statement in his Chapter IX.2, where he says that the retrograde arcs are inconsistent with the simpler models, and that it was Hipparchus who established this.

When Ptolemy comes to explain why he introduces the equant in his Chapter IX.5, he does so "in vague and insufficient terms", as *Pedersen* [1974, p. 268] justifiably remarks. He makes some confusing remarks about the times of maximum, mean, and minimum velocities. He also makes some confusing remarks about arcs on the ecliptic taken at the same phases. It is possible that these arcs are the retrograde arcs, but I think it more likely that they are the arcs travelled by a planet between recurrences of the same synodic phenomenon, such as its passage through the first turning point.

Whatever may be the explanation of Ptolemy's writing on this point, it does not sound to me like the writing of a genius who has invented the equant model and who has demonstrated that it "saves the phenomena" better than any other model known. It sounds like the writing of a person who is discussing an idea that is not his own and that he does not understand very well.[6]

In summary, by Ptolemy's own account, it is not correct that Ptolemy was the first person to make a systematic study of planetary motions. Hipparchus was the first person to do so, according to Ptolemy himself, but Hipparchus did not use the equant model, still according to Ptolemy. He used nothing more complicated than the model of an epicycle moving along an eccentric deferent. We do not know who invented the equant model, and we may not assume that Ptolemy did so simply because he claims the credit. He has made too many false statements

[5]Note that opposition to the sun, which can happen only for an outer planet, always comes in the middle of the retrograde arc. As Pedersen suggests, it may be this fact which led ancient astronomers to realize that the sun is involved in the motion of the outer planets as well as of the inner planets.

[6]Further, Ptolemy never establishes that the equant model is superior to other models. He merely asserts that it is.

about his work and the work of others to let us accept his statements as valid evidence. Still, we must admit the possibility that he did invent it.

Even if he did invent the equant model, though, he did not use it very well. I doubt that he ever explored the proper ratio that the two eccentrics in the equant model should have to each other. He takes them to be equal in all cases, but they should be unequal in all cases.

3. A Pedagogical Motivation for Fabrication, part (a)

Harrington [1978] and *Moesgaard* [1976 and 1980] have advanced what I have called the "honest pedagogue" theory [*Newton*, 1978]. However, they advance the theory from somewhat different points of view, and hence I shall discuss their explanations of the theory in different sections of the present chapter. Since it is somewhat simpler to discuss Harrington's viewpoint, I shall take it up in this section and take up Moesgaard's in the next.

Harrington advances the theory in these words: "If Ptolemy had a large body of poor data, he may have found, by trial and error, parameters for his models that were satisfactory in an average sense. But how could he possibly present this in the *Syntaxis*? One solution would be to invent 'observations' by calculation from the theoretical models . . . so that for clarity of exposition he could demonstrate the deduction of the parameters from the fewest necessary observations."

I have already [*Newton*, 1978] discussed this formulation in some detail, and I shall only summarize my comments. I made five main points:

1. In the earlier chapters of this work, I have given a fairly detailed discussion of the way in which Ptolemy deduced his parameters from observation. In no case does he deduce the complete set of parameters for any model from the fewest possible observations. He comes closest with the outer planets, for each of which he deduces four parameters from six observations. Of course it is true that he deduces some individual parameters in almost every model from the minimum number of observations.
2. Ptolemy often presents multiple determinations of parameters from multiple data. It would not have prolonged the *Syntaxis* unduly if he had presented the other determinations that the honest pedagogue theory requires him to have made. He would not have had to work through the mathematical details every time. It would have been sufficient to go through the method in detail with one set of observations. After that, he would have had only to present the new observations and say something like

this: "If we proceed with these data in the same way as before, we get the following results."

Incidentally, when Ptolemy makes multiple determinations of one or more parameters in a model, they always agree to an incredible level of precision. Thus, instead of fabricating the minimum set of observations needed to find his parameters, he has often fabricated a highly redundant set of observations.

3. The parameters that Ptolemy found are, by and large, not "satisfactory in an average sense." If he had in fact found parameters that agreed satisfactorily with a large body of poor but genuine data, they would have been much better than they are. Thus the only explanation that is consistent with the honest pedagogue theory is not that he used a large body of poor but genuine data but that, on the contrary, he selected a small body of data that gave consistent but poor results.
4. This explanation does not work either. In many cases it is not conceivable that there existed enough observations bad enough to result in Ptolemy's parameters. An example, but not the extreme one, concerns the apogee of Mercury. I shall return to this point in Section IX.6. There I shall show that, in order to have a reasonable probability of finding data bad enough to yield his value of apogee, Ptolemy would have had to have access to a systematic series of observations made over a time period equal to about 10^4 times the estimated age of the universe. If Ptolemy had such a body of data, it would probably have taken him at least as long to reduce them and to find a set of data that would yield his value of apogee.

   Even if he did find his value occurring a few times in an enormous set of solutions, why did he choose it over values that would have occurred an overwhelming number of times?
5. Often there is no pedagogical advantage in what Ptolemy did. A good example concerns the star catalogue. Ptolemy fabricated the longitudes in his catalogue by adding $2° 40'$ to the longitudes in a catalogue originally prepared by Hipparchus. He then pretended that he had measured these longitudes, and he even described the method that he claimed to have used to make the measurements. I see no pedagogical value in his claim to have made the observations. However, there would have been considerable value in explaining what he actually did and in explaining that this is a valid way to "update" a star catalogue.

To summarize this variant of the honest pedagogue theory, the idea is that Ptolemy felt that it would be unbearably tedious for his readers if he showed them how he actually found his parameters. Hence, to spare them this boredom, he fabricated a few observations and pretended to derive his parameters from them. However, this does not explain what happened in the *Syntaxis*, as I have just shown.

4. A Pedagogical Motivation for Fabrication, part (b)

Now we turn to the variant of the honest pedagogue theory that Moesgaard has advanced, which is rather more complex in its argument. I have already alluded to Moesgaard's work in Section II.7 in connection with the obliquity of the ecliptic. He has presented his theory in two main papers [*Moesgaard*, 1976 and 1980], and I believe that the following expresses his theory correctly: Over a period of many centuries, Babylonian astronomers had developed accurate numerical schemes for making astronomical calculations, but they never developed any models or theories to correspond, so theirs remained a purely numerical astronomy. When Greek astronomers did begin to develop models and theories, they wanted to preserve continuity with the highly successful Babylonian astronomy. To do this, they either took over Babylonian parameters where appropriate or derived the parameters in their models by fitting to the Babylonian schemes. Thus, as *Moesgaard* [1980, p. 95] writes: ". . . I find Ptolemy trapped in the dilemma between, on one hand, the ideal of initiating a new doctrine of astronomy based on fresh data, and on the other hand, the necessity of preserving the continuity with parameters which had worked well for centuries and had, therefore, deservedly grown canonical."

I have already discussed in Section II.7 how Moesgaard derives the famous obliquity of 23;51,20 degrees by Babylonian methods. He has also studied the derivation of other solar parameters by Babylonian methods based upon the use of lunar eclipses rather than equinox and solstice data. For example, he finds [*Moesgaard*, 1980, p. 91] that the maximum equation of the center for the sun is either 2;34 or 2;18 degrees, depending upon how he derives it from Babylonian results. Hipparchus's value for the eccentric, which Ptolemy adopted without change, is 2;30/60, which leads to a maximum equation of 2;23 degrees. Moesgaard considers this to be adequate agreement, since Hipparchus might have used a different set of lunar eclipses.

It seems to me that there are serious difficulties with this variant of the honest pedagogue theory, just as there were with Harrington's, and I have discussed some of them earlier [*Newton*, 1977a]. In discussing these difficulties, I shall use the sun as an example, since the solar theory has been the main thrust of Moesgaard's research (that is, of his research that is relevant to the present work).

Before taking up these difficulties, we need a little background on the general nature of Babylonian astronomy beyond the fact that it was purely numerical. *Neugebauer* [1955, volume 1, pp. 41ff] gives an extensive description of the matter, but we need only a few details. To start with, in Babylonian astronomy as in Greek astronomy and modern astronomy, calculating the position of a body is divided into two parts.[7] First we calculate the mean position, that is, the position of a body moving with uniform angular velocity. Then we calculate the equation of the center, which is the true position minus the mean position. Finally, of course, we add the equation of the center to the mean position. The equation of the center is the sum of various periodic functions.

There were two main systems of Babylonian astronomy, which differed in the nature of the periodic functions that they used; they did not have sinusoidal functions. In what we call System A, they used periodic step functions, and in System B they used periodic zigzag functions. The former are discontinuous functions. The latter are continuous functions made up of straight line segments. With this information, we can turn our attention back to the honest pedagogue theory.

First, I do not see why Ptolemy should have been trapped between the introduction of new data and maintaining continuity with Babylonian astronomy. To me, the distinctive features of Babylonian astronomy are its recognition that the difference between the true and mean positions can be built up from periodic functions, and the ways in which it performed the necessary calculations. Calculation of the mean position is a rather trivial matter, once a mean motion or a mean period has been determined.

Thus I do not see why Ptolemy should have felt constrained to preserve a length of the year that maintained continuity with the Babylonians. He clearly states [*Syntaxis*, Chapter III.1] that the length of the year should be based upon a time span of data that is as long as possible. Thus he should not have been disconcerted if observations made by him did not agree with the Babylonian value for the year; it should have seemed an almost necessary consequence of the fact that he had a longer time base available to him. In other words, he had no excuse that I can see for adopting a received value for the length of the year.

Mean motions should not become canonical, and Ptolemy did not take them to be so. We saw in Chapter III that he did not hesitate to alter the received values of γ' and ψ' for the moon, although he did adopt the Babylonian value for the length of the synodic month. Thus there is no reason why he should have accepted a particular length of the year as canonical.

[7]This is not true of very modern astronomy. The trend today seems to be toward solving equations of motion by numerical integration instead of by perturbation theory.

Second, Ptolemy, and Hipparchus before him, did not adopt the Babylonian length of the year. It would be surprising if they did: Ptolemy and Hipparchus knew of the precession of the equinoxes, and thus they knew that there is a difference between the tropical and sidereal years, but the Babylonians did not know of this distinction. In fact, the Babylonians used the same length of the year in both Systems A and B, and this value is [*Neugebauer*, 1955, volume 1, pp. 45 and 70] 12;22,8 mean synodic months. The Babylonian value for the month is (Section III.3) 29;31,50,8,20 days. Thus the Babylonian value for the year is $365^d.260\ 638$, in decimal form. This is the sidereal rather than the tropical year.

Now we know that Ptolemy fabricated his solar observations by using Hipparchus's length of the year, not the Babylonian value.[8] Thus he did not fabricate the data in order to preserve continuity with Babylonian astronomy, but with older Greek astronomy that specifically differed with the Babylonian. More precisely, as we saw in Chapter II, he fabricated his solar data in the way that cost him the minimum amount of effort.

Third, this version of the honest pedagogue theory puts Hipparchus in a rather odd light. Certainly the simplest way to find the parameters of the solar orbit is to use measurements of the equinoxes and solstices. We have strong evidence that Hipparchus made extensive measurements of the equinoxes and solstices, and the parameters in his (and Ptolemy's) solar theory are consistent with these data. On the other hand, according to Moesgaard, Hipparchus determined his parameters by means of Babylonian sources. If he did so, it would look strongly as if he fabricated the equinox and solstice data that are attributed to him. Thus the honest pedagogue theory seems to be an indictment of Hipparchus rather than an exculpation of Ptolemy.

Fourth, the honest pedagogue theory cannot explain many of Ptolemy's actions. It does not touch upon his fabrication of the lunar parallax at the quarter moons (Section IV.4), nor upon his fabrication of the precession of the equinoxes (Section V.1), nor upon his fabrication of the apogee of Mercury (Section VII.2), to give but a few examples. There could have been no way to preserve continuity with Babylonian astronomy for at least the first two of these topics because they were unknown in Babylonian astronomy.

Finally, even if Ptolemy were caught in the dilemma that Moesgaard postulates, I do not see why this would justify his fabrication of data. I can see why he might wish to preserve continuity with Babylonian astronomy even after he found out that it conflicted with new observa-

[8]As I understand him, Moesgaard concurs that Ptolemy fabricated his solar data. He disagrees with me only in the motivation for this fabrication.

tions. I might even bring myself to accept that he could legitimately have omitted any mention of this conflict from his treatise on astronomy, although such conduct would be considered improper today, and I would have to think about the matter more carefully before accepting it. But I do not see why he would be justified in fabricating data to make it appear that the data agreed with the received astronomy when they did not. That is unacceptable conduct by the standards of either his time or ours.

I see no way to escape the conclusion that I have stated several times before: Ptolemy fabricated data for the purpose of demonstrating that his ideas agreed with observation, even though in at least a number of instances he knew that they did not. This was deliberate intellectual fraud.

5. A Theory of Errors

In judging what Ptolemy did, from the standpoint of both his competence and his integrity, we should of course judge him by the standards of his own time and by the intellectual, scientific, and mathematical equipment that was available to him. I was careful to do this in *CCP*. However, when I wrote *CCP*, I did not realize that there are several important factual errors in what we may call the "standard" literature on Greek astronomy, and I shall take up some of the most important of these errors in the next few sections.

It is not news that the standard literature on ancient astronomy contains a number of errors. *Dreyer* [1905], in the preface to his *History of the the Planetary Systems from Thales to Kepler*, points out that ". . . a number of legends on subjects connected with the history of the cosmical systems have been repeated time after time, not only in works on the general history of science and literature . . . but also in books dealing specially with astronomy. Among errors long ago refuted but still frequently produced in print may be mentioned: that Thales knew the earth to be a sphere; that Pythagoras and his school taught the motion of the earth round the sun; that Plato taught the daily rotation of the earth and in his old age inclined to the heliocentric system; that the Egyptians knew that Mercury and Venus move round the sun; that the lunar variation was discovered by Abu'l Wefa; that King Alfonso the X of Castille found the orbit of Mercury to be an ellipse (which has not been refuted before, so far as I know),[9] and that Cusa and Regiomontanus anticipated Copernicus." Dreyer also says that the opportunity to correct these and other errors was ". . . what chiefly induced me to write this book, . . ."

[9]This parenthesis is Dreyer's.

Unfortunately a number of recent writers about Ptolemy have based their conclusions upon the standard literature without realizing that it contains important factual errors. For example, *Gingerich* [1980] wrote about Ptolemy: "Since he had no theory of errors, he could cope with the multiplicity of data by including in the *Almagest* only the minimum number of observations required to determine the parameters." In an unsigned news story in *Scientific American* [1979], in the section called "Science and the Citizen" in the issue for March, 1979, we find a similar statement:[10] "It was not until the development of statistical and probabilistic methods at the end of the 18th century that natural philosophers started to report all random observations, because only then did they have the techniques for interpreting large bodies of imprecise data. . . ." This is like saying that scientists could not deal with velocity before the invention of the differential calculus.

In considering briefly the matter of publishing all random observations, we must naturally confine ourselves to astronomers whose work has survived in sufficient quantity to let us make a judgment. We probably cannot say of any individual astronomer that he published *all* his observations, but that is not the real question. The real question is: Did he publish redundant observations? That is, did he publish appreciably more observations than the minimum needed to determine his parameters?

With the understanding just stated, we can say that *every* known astronomer, ancient, medieval, or modern, published redundant observations, and I shall give three examples. *Dreyer* [1926, volumes 10-13] has published four volumes of observations made by Tycho Brahe or by others in his observatory. I estimate that there are of the order of 10^4 observations in these four volumes, and they may in fact include all of Tycho's observations. *ibn Yunis* [1008] published hundreds of observations made by himself and other astronomers, dealing with the lengths of the year and of the seasons, with eclipses, and with the positions of the moon and planets.

The final example is Ptolemy himself. He published redundant data and made multiple determinations of at least some of the parameters in each model he dealt with. The extreme number of determinations concerns the precession of the equinoxes. Here he made ten independent determinations using three independent methods.

Sometimes when he makes multiple determinations of a parameter, he gets exactly the same value every time. At other times, though, he gets slightly different values, presumably because of rounding problems in his calculations and particularly in rounding the "observational data" so

[10]The statement is attributed to "Victor E. Thoren of Indiana University and Owen J. Gingerich of Harvard."

that they would look plausible. When he does the latter, he then has the problem of choosing his final value, and in doing so, he shows us what his theory of errors was: When one is confronted with several different values for a quantity, one chooses a value somewhere between the smallest and the largest. Apparently he never took the formal average in the modern sense, although his choice is often fairly close to the average. His procedure, though rudimentary, is a genuine theory of errors.

We should note that Harrington's form of the honest pedagogue theory specifically requires Ptolemy to have a theory of errors. We should also note that Greek astronomers were well aware of various types of error and of their effect on the parameters inferred from them. In perhaps the most elaborate example known, Theon of Alexandria examines the effect that a bias of 6 minutes of arc in measuring the meridian zenith distance would have on the inferred declination and longitude of the sun [*Pedersen*, 1974, p. 148]. Further, both Ptolemy and Hipparchus showed an understanding of the effects of observational error upon their result. In sum, we see that Ptolemy had a theory of errors in both senses of the term.

6. Ptolemy's Selection of Data

As an explanation of the fact that Ptolemy's data fit his theories so well even though they are seriously in error, *Gingerich* [1980] suggests the following explanation: "The fact that . . . erroneous observations fit the theory so well would then be explained by his use of carefully selected data . . ., and not by fudging the observations." Many others have made the same suggestion. For example, *Goldstein* [1978] writes, after outlining a certain argument about Ptolemy: ". . . this suggests that those [11] displayed were chosen for their agreement with a theory that in some sense derives from the larger body of observations." As a final example, *Swerdlow* [1979, p. 530] writes: ". . . there is no doubt that Ptolemy was in fact selecting from some substantial number of observations analyzed in preliminary work not reported in the *Almagest*, and that such prior analysis contributed to determining both his choice and his use of particular observations."

It seems to me that these quotations imply that Ptolemy had a theory of errors, but one that is different from the theory I outlined in the preceding section. What these writers are saying, as I read them, is the following: When he was confronted with a distribution of values, Ptolemy chose a value from one tail or the other of the distribution.

[11]That is, those observations that Goldstein has just been discussing.

For example, consider Ptolemy's equinox measurements. His three measurements are all too late. The two autumnal equinoxes are too late by about 34 hours and the single vernal equinox is too late by about 21 hours. The standard deviation in measuring an equinox that was achieved by Greek astronomers earlier than Ptolemy is [*CCP*, Section V.2] about 7 hours, and there is no reason why Ptolemy should have done any worse using the methods he describes [*Syntaxis*, Chapter I.12]. If anything, we would expect an improvement in techniques between earlier astronomers and his time, and hence we should expect greater accuracy from him.

Further, there are no serious sources of bias in the methods he describes, and there are no sources at all that would bias both autumnal and vernal equinoxes in the same direction. Thus, if Ptolemy made many measurements of the equinoxes, or if he had available many measurements that had been made between Hipparchus and himself, the data would have been centered near the correct value, with most values having errors of less than 7 hours. Yet, according to the idea expressed above, Ptolemy chose to use a small set of three values lying, in modern terms, about 3 standard deviations away from the center for the vernal equinox and 5 standard deviations away for the two autumnal ones.

In other words, what the writers just quoted are saying is equivalent to the following: In ancient Greek times it was accepted practice to adopt a small set of values at the tail end of a distribution while ignoring the vast bulk of the data that lie near the center. So far as I know, ancient Greek writers are silent on this point, but I see no reason to assume that any of them would have thought that this was a sensible thing to do. I suspect that they are silent on the point because it never occurred to them that anyone would do such a thing.

However, there is another important point besides the point of whether ancient Greek astronomers thought it was sound practice to adopt the tail of a collection of values while ignoring the vast bulk of the data. This is the question of whether a set of observations existed from which Ptolemy could have made his selection. I have already alluded to this matter in Section IX.3.

Let us take the apogee of Mercury as an example. We saw in Section VII.2 that Ptolemy started by finding apogee from observations 1, 2, and 3 in Table VII.1, and that he found the value 185 5/6 degrees. We should remember that these observations are probably genuine, but that Ptolemy made a serious mistake in analyzing them; he should have gotten about 215° or so. He then used observations 4, 5, and 6 and found 186° exactly.

Since the first determination is probably based upon genuine data, it poses no problem. The problem comes from the second determination, which agrees accurately with the first determination and which agrees exactly with Ptolemy's final theory of the apogee, within the level of rounding used. This level is 5′ of arc, so the second determination lies within a preassigned interval of width 10′, but which is in error by nearly 30°.

In observation 4, the longitude L_S of the mean sun was 147;50 degrees and the maximum elongation E was 21;40 degrees. Ptolemy did not have an observation in which W was 21;40, so he interpolated between observations 5 and 6 to find that L_S is 224;10 when W is 21;40. For brevity, let me call this interpolated result observation 5/6. The most conservative course we can take at this point is to assume that observations 5 and 6 are genuine. This does not mean that "observation 5/6" is genuine, however, as I shall show in a moment.

The fact that L_S lies in an interval of width 10′ imposes exceedingly tight limits on the measured longitude of Mercury. For conditions near those of observation 4, I find that changing E by 1° requires changing L_S by about 18°. Hence, if L_S lies within a preassigned interval of 10′, the elongation and hence the measured longitude must lie within an interval whose width is 10/18 minutes of arc. By calculating the longitude of Mercury from modern theory, I find that the center of the interval corresponds to an error of 1°.80.

Let us suppose that the standard deviation in measuring the longitude of Mercury is the same value that we find in measuring the position of a star. This is [*CCP*, p.216] about 21′ in each coordinate. If we use this value, the width of the interval is 0.0265 standard deviations and the center of the interval has an error of 5.14 standard deviations. If we assume that the normal law of error applies at these rather extreme conditions, the probability that a genuine error could meet these conditions is 1.93×10^{-8}.

Now observations 1 through 6 lie within a time span of 25 years. Suppose that Ptolemy was willing to use observations lying within 25 years in either direction from −261. Then the total time span he was willing to use was 50 years. If we assume that there was one useful set of observations per year,[12] the data sample available to Ptolemy had 50 sets. The probability that this set of observations would have contained an observation meeting the conditions placed on observation 4 is about $50 \times 1.93 \times 10^{-8}$, which is close to 10^{-6}.

As an aside, the reader may wonder how Ptolemy got the longitude of Mercury to lie in an interval only 10/18 minutes wide when he reported it only to the nearest multiple of 10′. The answer comes from

[12] The synodic period of Mercury is about 116 days, so in principle there are about 3 opportunities per year to make the necessary observations. However, it is unlikely that any ancient astronomers observed the maximum elongations of Mercury systematically enough to have recorded all the possibilities. In this necessarily approximate discussion, I think it is sufficient, if not optimistic, to assume one set of observations per year.

the fact that "observation 5/6" is the result of an interpolation. After he found observation 4, whether he found it by fabrication or by selection from a large sample, he simply misstated the result of the interpolation so that it seemed to give exact agreement. Specifically, he said that the interpolation gives 224;10 degrees as the value of L_S, but the correct value, rounded to the nearest multiple of 10′, is 223;50 degrees. Thus observation 5/6 is fabricated even though the individual observations 5 and 6 may be genuine.

Now let us look at the four observations in Table VII.1 that Ptolemy claims to have made himself; these are observations 7 though 10. Using these, he obtained the values 189;52,30 and 190;15 degrees for the position of apogee. Here the agreement with Ptolemy's theory is not as close as it was for the value obtained from observations 4, 5, and 6, although we should note that the errors are carefully balanced to lie on either side of the value 190°, which is Ptolemy's final choice for the value in his own time. In analyzing these observations, let us take the width of the interval in which L_S lies as 20′ rather than 10′. In order to be conservative, let us also take for granted the values of E in this set of four observations and ask for the probability that Ptolemy's data set could have contained matching values of W.[13]

The errors in the W's are almost exactly 3.0 and 3.5 standard deviations, and the width of the interval is 0.053 standard deviations. The probabilities that genuine observations could meet these conditions are 0.000 235 and 0.000 046 3, respectively.

Next let us look at the number of suitable observations that might have been found in the collection of Ptolemy's own observations. It is certainly peculiar that he found two sets of E's and W's that exactly matched each other in only nine years of observations even though he could not find any matches at all in the old observations. I do not believe that this exact matching could have happened by chance more than these two times in Ptolemy's working lifetime as an astronomer, so he had no opportunity to select these data from a large set.

The probability that Ptolemy's data set could have contained observations 4, 8, and 10 is the product of the three probabilities we have found, which is almost exactly 10^{-14}.

In other words, Ptolemy had no reasonable chance of selecting the data he used from a larger set of genuine observations unless the set contained something like 10^{14} observations. At one observation per year, it would take 10^{14} years for astronomers to amass the observations needed. This is of the order of 10^4 times the estimated age of the universe.

As I have said before, I do not mean for probabilities like 10^{-14} to be taken literally. All I mean is that the probabilities are far beyond the

[13]The errors in the W's are less than the errors in the E's [*CCP*, Table X.1].

realm of ordinary experience. There are at least two reasons why we should not take them literally.

In the first place, we do not know accurately the values that we use in calculating the probabilities. In particular, we do not know the standard deviation accurately. I used 21′ for the standard deviation, but I cannot say that it is really 21′ rather than, say, 18′. Changing the standard deviation from 21′ to 18′ would make a large change in the exponent in the probability.

Second, there is the question of whether the normal law of error applies in these extreme conditions. In some cases, I suspect that it does not, and that its use gives too large a probability. Consider Ptolemy's measurements of the equinox again. If we assume for the moment that his two observations of the autumnal equinox are genuine, there must have been a bias in his meridian circle which caused him to put both of them about 34 hours late. Since the declination of the sun changes by 1′ per hour near an equinox, the direction that Ptolemy called the equator must have "leaned" too far south by 34′. The same bias in direction would apply to a vernal equinox, so the event that he called the vernal equinox would come 34 hours before the correct time. Since Ptolemy observed it to come 21 hours late, he made an accidental error of 55 hours in his observation, on top of his bias, and hence he made an error of 55′ in judging the position of the center of the sun.

The standard deviation of 7 hours that I used includes an allowance for all sources of bias as well as for the accidental error in a single observation. In considering the error of 55 hours, we should use only the accidental component, which would have a standard deviation of perhaps 3 hours rather than 7 hours. However, to be conservative, let us continue to use 7 hours. Then the error is 55/7 standard deviations, and the probability that an error can be this large is about 2×10^{-15}.

Now let us look at Ptolemy's claimed method in more detail. He put a pin at the center of a graduated circle, so that it cast a shadow onto the circle at noon. He moved a slide along the circle until it was symmetrically located with respect to the shadow. Then, at his leisure, he could read the position of the slide; he did not have to read the position of the shadow "on the fly". The half-width of the shadow was about 15′, so the point that Ptolemy called the center of the shadow lay entirely outside it by 40′, almost three times its half-width.

I think we may say flatly that such an error is impossible. That is, the probability calculated by the normal law, which is 2×10^{-15}, is too large. The true probability is 0.

I think the same thing applies to Ptolemy's alleged measurement of the lunar parallax. As I said in Section IV.4, Ptolemy could not have seen any part of the moon in the field of view of his instrument if he had really done what he claimed to do on this occasion. I think the error he made is impossible for a genuine observation, and thus it was impossible for him to have selected this observation from a body of genuine data.

7. The Role of Observation in Ancient Greek Astronomy

There is another error in the standard literature on Greek astronomy that I first encountered in discussions of Aristarchus and his determination of the sizes of the sun and moon. *Aristarchus* [ca. −280] uses several observational quantities in this determination, and one of them is the apparent diameter of the sun and moon (he takes the two diameters to be equal). He takes this diameter to be 2° rather than the correct value of about 0°.5. It does not seem likely that this value came from any serious attempt to measure the diameters, so *Pannekoek* [1961, p.120] remarks that Greek astronomers were primarily geometers who happened to pick out the heavens as a source of interesting problems. "Hence," he writes, "the astronomical quantities were treated somewhat superficially, their precise values did not matter; ingenuity was exhibited in the solution of the geometrical problem."

Before going on with the main theme of this section, let me digress a moment to point out what Aristarchus actually did; I have discussed his treatment of the sizes in considerable detail elsewhere [*CCP*, Appendix B]. He estimated the sizes of the sun and moon but not the distance of either. He used the apparent diameter of the bodies only in order to show that a certain quantity in his calculations was negligible, and he showed that this was so even if the diameter was as large as 2°.[14] Thus taking the diameter to be 2° did not degrade the accuracy of his work.[15]

In recent years this idea (that Greek astronomers were not concerned with the physical accuracy of their work, but only with the ingenuity of their mathematics) has often been advanced as a justification for

[14]Aristarchus does say in his proposition 11 that the distance of the moon lies between 45/2 and 30/1 times the diameter of the moon, but he says this before he has found the lunar diameter, and he uses it only in the process of justifying the neglect of the quantity just mentioned. He does not return to it after he finds the lunar diameter.

[15]Most discussions of his work say that he found both the sizes and distances of the sun and moon, and that his value of the diameter introduced a serious error (a factor of about 4) into the distances. However, Aristarchus was careful not to do anything whose accuracy was affected by using 2° for the diameter, and in particular he did not calculate the solar and lunar distances (except in the minor way mentioned in the preceding footnote). This means, in my opinion, that Aristarchus *was* concerned with the physical accuracy of his results, and that he was not merely interested in the clever solution of a difficult geometrical problem.

The title of his work is often given in English as *On the Sizes and Distances of the Sun and Moon*. Since Aristarchus did not deal with the distance of either body, I doubt that he gave the work this title (or its equivalent in Greek); I suspect that it has been invented by a later editor or translator. Accordingly I give the title as *On the Sizes of the Sun and Moon* in the list of references for this work.

Ptolemy's fabrication of data. For example, we see the idea attributed to Thoren and Gingerich in the news story in *Scientific American* that I mentioned in Section IX.5: "The people who did astronomy in Ptolemy's time were mathematicians, who concerned themselves with proof, rigor, logic and consistency rather than with observational accuracy."

Gingerich [1980] expresses a related idea when he writes: ". . . I suspect that Ptolemy, convinced in [*sic*] the intrinsic soundness of his theory, simply replaced the only partially trustworthy observations by what he perceived to the 'correct' data.[16] Thus Ptolemy, like many of the brilliant theoreticians who followed him, was perfectly willing to believe that his theory represented Nature better than the error-marred individual observations of the day. As one of America's Nobel laureates remarked to me, any good physicist would do the same today."[17] *Neugebauer* [1957, p.185] says much the same thing: "The ancient astronomers rightly had greater confidence in the accuracy of their mathematical theory than in their instruments."

Incidentally, the Nobel laureate must have made his remark in a different context. No reputable physicist today or in any day would "replace" observations by fabricated data. If a physicist were discovered doing so, he would be drummed out of the profession.[18]

Also incidentally, ancient Greek astronomers would have been wrong if they had had "greater confidence in the accuracy of their mathematical theory than in their instruments." In the specific context, Neugebauer is saying that a longitude calculated from Ptolemy's lunar theory is more accurate than a Greek measurement of a stellar longitude. Yet the standard deviation of the error in his lunar theory is [*CCP*, Table XIII.3] $0°.58$ ($35'$), exclusive of the bias in longitude, while the standard deviation of a Greek measurement of a stellar longitude is [*CCP*, Table IX.1] about $22'$. "Their instruments" were in fact more accurate than their theories.

As I understand the argument being made, agreement of mathematical astronomy with observation was considered to be unimportant by the ancient Greek astronomers, so the fact that Ptolemy made up his data is also unimportant; it was completely in accord with the standards of the time. Thus, as the argument goes, Ptolemy's fabrication

[16]The quotation marks are Gingerich's. In the original, a new paragraph begins at this point.

[17]*Hartner* [1977], however, in his discussion of the role of observation in ancient and medieval astronomy, believes that this attitude prevailed ". . . throughout the ages" down to about the time of Galileo, but that it has now vanished from science, if I understand him correctly.

[18]Further, most advances in the sciences have come from exploring the differences between "error-marred" observations and theory. This was as true in ancient times as it is today.

of his data does not provide any justification for concluding that he acted in bad faith.

In this section I shall study the attitude of several Greek astronomers toward the relation between observation and theory. I shall do this by investigating what they actually did in the matter of observation and, where possible, by considering what they actually wrote on the matter.

a. PLATO. Plato was a philosopher rather than an astronomer, but he wrote quite a bit about astronomy from his own philosophical point of view. I suspect that many historians of science, when they are writing about the Greek attitude toward physical observations, are unconsciously influenced by Plato's attitude. Thus it is appropriate to start this study with Plato even though he was not himself an astronomer.

Plato was greatly concerned with the questions of what true knowledge is (should *true* perhaps be in quotation marks?) and how it is to be acquired. His thought on the subject, as I understand it, is succinctly expressed in the dialogue called *Phaedo* [*Plato*, ca. −380, p. 53 in the cited translation], which is a possibly imaginary conversation between Socrates and some of his friends on the day of his execution. In this dialogue, Plato has Socrates say: "What again shall we say of the actual acquirement of knowledge?—is the body, if invited to share in the inquiry, a hinderer or a helper? I mean to say, have sight and hearing any truth in them? Are they not, as the poets are always telling us, inaccurate witnesses? . . . Then when does the soul attain truth?—for in attempting to consider anything in company with the body she is obviously deceived. . . . Then must not existence be revealed to her in thought, if at all?"

Heath [1913, Chapter XV] gives an excellent analysis of the writings of Plato that pertain to astronomy. He puts the essence of Plato's thoughts on astronomy into these words [p. 138]: "There is no doubt that Plato distinguishes two astronomies, the apparent and the real, the apparent being related to the real in exactly the same way as practical (apparent) geometry which works with diagrams is related to the real geometry." Thus, for example, a "true" circle is a line with no width, whose every point is equidistant from the center. The "circle" that a geometer draws with a compass to illustrate a theorem, which is the only kind of circle that can be perceived by the senses, is obviously but an imperfect representation of the "true" circle that can be perceived only by the mind.

Similarly, in Plato's thought, there is a "true" universe of astronomy that can be perceived only by the mind, and the

> universe that we perceive by our senses is only an imperfect representation of it. It follows, among other things, that we should not be concerned by imperfections in the visible universe, and we should not spend our energies in trying to explain them.
>
> Heath goes on to ask: ". . . what form would astronomical investigations on Plato's lines have taken in actual operation?" Specifically, he asks how Plato would have reacted to the minor perturbations of Uranus that led to the discovery of Neptune, and he answers in these words: "For Plato says that the person who thought that the heavenly bodies should always move precisely in the same way and show no aberrations whatever would properly be thought 'absurd',[19] and that it would be absurd to exhaust oneself in efforts to make out the truth about them; hence, on this showing, the visible perturbations of Uranus would scarcely have seemed to Plato very extraordinary or worth any very deep investigation. . . ."
>
> It seems to me that these ideas of Plato lead directly to the idea that "the people who did astronomy in Ptolemy's time were mathematicians, who concerned themselves with proof, rigor, logic and consistency", which are matters of the "soul", "rather than with observational accuracy", which is a matter of the body. However, we need to ask: Were Greek astronomers true Platonists, or did they proceed in practice by insisting that their theoretical constructs had to agree with observation?

Before going on to consider other Greek writers, I wish to add some further remarks in fairness to Plato. As Heath says, it is quite possible that Plato, in the views that have just been summarized, is consciously exaggerating for the purpose of making a point. Indeed, we must admit that idealization of the visible plays an important part in astronomy and in all the other sciences.[20] Further, as Heath points out, Plato modified his attitude toward observation in other writings that are probably later. *Heath* [p. 140] quotes another dialogue in which Plato writes: "Sight, according to my judgement, has been the cause of the greatest blessing to us, inasmuch as of our present discourse concerning the universe not one word would have been uttered had we never seen the stars and the sun and the heavens . . . whence we have derived philosophy, than which no greater good has come. . . ."

[19]The quotation marks are Heath's.

[20]However, I think there is an important difference between much modern idealization and that of Plato. For example, in developing our theories of the planetary motions, we idealize a planet as a point mass. We do not, though, take this idealization as the "true" planet. Instead, we regard it only as an approximation made in order to simplify the analysis.

b. METON AND EUCTEMON. Histories of Greek astronomy often start with Thales of Miletus, and he is often called the father of Greek astronomy. However, it is difficult to separate fact from legend with regard to him,[21] so let us go on to Meton and Euctemon. Ptolemy [Chapter III.1] refers to a measurement of the summer solstice made by these astronomers (or perhaps by their students) in −431. We also know that a new Athenian calendar (see *CCP*, p.94ff, for example) was devised in or about this time. Some sources attribute the calendar to Meton, some to Euctemon, and some to both. The length of the year in this calendar was 365 + (1/4) + (1/76) days. Euctemon also measured the lengths of the seasons [*Geminus*, ca. −100] with results that are rather accurate; see my discussion of them in *CCP* [pp. 80ff].

It is clear from what they did that accuracy, agreement with observation, was the central goal of Meton and Euctemon. Further, the calendar mentioned was an elaborate mathematical device that reconciled the solar and lunar periods to an accuracy of the order of about 10 minutes per year. Some may object that Meton and Euctemon were not astronomers but mere calendarians, but it seems likely that they were pursuing astronomy as well as they could with the means available.

c. CALLIPPUS. Callippus, about a century after Meton and Euctemon, also made careful measurements of the seasons and of the length of the year. He improved the calendar of Meton and Euctemon by omitting one day every 76 years, and thus he became, so far as I know, the first astronomer to adopt the "Julian year" of 365¼ days. Clearly he also was interested in observational accuracy, but perhaps some will dismiss him as another calendarian.

d. EUDOXUS. Eudoxus was slightly earlier than Callippus.[22] He wrote a book about the appearance of the heavens, which the

[21]One legend about Thales that still continues is that he predicted the "very year" in which there was a total solar eclipse visible in Asia Minor, and that the eclipse was that of −584 May 28. However, the source of the legend [*Herodotus*, ca. −446, Book 1] does not say that the event described was an eclipse, and it may have been something else. Even if we assume that it was an eclipse, the date could range at least from −607 to −581 [*Newton*, 1970, pp. 94ff] . Thales cannot be dated accurately by means of the "eclipse", but this has not kept many historians from claiming that they have done so.

[22]*Dreyer* [1905, p. 88] says that Eudoxus was the first astronomer to propose a year of 365¼ days, but he does not identify any sources by which one can check the statement and I have not tried to pursue the question. In any case, an early (−4th century) calendric system that embodied a year of 365¼ days was called the Callippic cycle.

poet Aratus, about a century later, either converted into poetic form or used as the basis for a poem. Still later, *Hipparchus* [ca. −135] wrote his *Commentary on Aratus and Eudoxus*, as it is often called; this is the only work of Hipparchus that has survived, except for quotations or paraphrases.[23] Eudoxus also developed a planetary theory usually called the theory of homocentric spheres, which *Dreyer* [1905, Chapter IV] describes in considerable detail.

Eudoxus's model involved a number of spheres, all concentric with the earth, that rotated about various axes and that carried the bodies in the solar system with them as they went. Since the poles of rotation of some spheres were carried on the surfaces of other spheres, the model was capable of producing motions that were quite complex as viewed from the earth at the center. *Dreyer* [1905, p. 87] calls this system "the first attempt to account for the more conspicuous irregularities" of planetary motion. It is clear that Eudoxus was attempting to derive a geometrical system that would agree with observation, not be indifferent to it. Further, it seems the system of Eudoxus was finally abandoned by all Greek astronomers because it failed to agree with observation in one important way. Its main defect was that it required a particular planet always to have the same brightness, which is clearly not the case. Observation, not ingenuity or mathematical rigor, was the touchstone by which Greek astronomers tested the system of Eudoxus and rejected it.[24]

e. HIPPARCHUS. As I have already said, the only surviving work of Hipparchus in its original form is his *In Arati et Eudoxi Phaenomena Commentariorum Libri Tres*, which is usually called *Commentary on Aratus and Eudoxus* in English. In this work, Hipparchus typically proceeds by quoting a passage from the earlier work which describes some quantitative feature of the heavens, such as the distance between two stars. He then presents his own measurement of the feature, which usually disagrees with the result given by Aratus and Eudoxus. *Neugebauer* [1957, p.185] says that the *Commentary* contains 471 spherical coordinates of stars, of which 64 are declinations, 67 are right ascensions, and 340 are other spherical arcs, but

[23] This statement does not apply to Hipparchus's star catalogue, which survives in a slightly altered form in the *Syntaxis*.

[24] In Section II.8 I quoted a passage which showed that annular eclipses were probably known to Greek astronomers at least two centuries before Ptolemy. The original purpose of the passage was to refute the system of Eudoxus.

none of which are either longitudes or latitudes. However, in a separate study [*Newton*, 1974], I found only 40 declinations; I did not try to count the other coordinates. In the *Commentary* it is clear that accurate measurement is "the name of the game"; stellar coordinates were not quantities whose "precise values did not matter. . . ."

During the many centuries between the discovery of precession and the discovery that the ecliptic is rotating, astronomers thought that the latitude of a star is a constant while its longitude increases at a uniform rate. Thus latitude and longitude seemed to them like the ideal stellar coordinates. The fact that Hipparchus did not use them in the *Commentary* suggests that he wrote this work before he discovered the precession. He did use them in his star catalogue, which still exists in the *Syntaxis* in the altered form that Ptolemy gave it, and this suggests that the catalogue is later than the *Commentary*. However this may be, Hipparchus did measure the coordinates of somewhat more than 1000 stars and presented them in an organized form in his star catalogue. It is again clear that he regarded accurate and organized measurements as an important, even a necessary, feature of astronomy.

In addition to his activities about the stars, Hipparchus made careful measurements of the length of the year, the lengths of the seasons, the deviation of the moon from the mean moon, and the positions of the planets. However, Hipparchus was not only a measurer. He developed models for the motion of the sun and moon, he developed methods of predicting eclipses, both lunar and solar, and he attempted to make an accurate determination of the solar distance. In addition, if Ptolemy is accurate in this matter, Hipparchus developed models for the motions of the planets[25] and showed that they did not fit the observations.

f. PTOLEMY. Ptolemy tells us directly the attitude that he professed to have about the relation between theory and experiment. He emphasizes that mathematical theories of astronomy must be based upon observation, and he writes in Chapter III.1 of the *Syntaxis*: "We think that it is right to explain the phenomena by the simplest possible hypotheses, provided that they do not conflict with the observations in any important way."[26]

[25] See the discussion in Section IX.2.

[26] As I have already noted in Section IV.2, at a certain point in his lunar theory, Ptolemy could have obtained a certain quantity by calculation, but he chose instead to deduce it from observation. His reason for doing so, he said, is that there is more surety in the results if one relies upon the observed phenomena.

> This passage in fact expresses two important aspects of what is still taken as the proper relation between theory and observation (or experiment). One is that a theory, to be regarded as valid, must agree with observation as well as we can expect. The other is that, as between two theories that are equally successful in agreeing with observation, we usually choose the simpler one, or, as is sometimes said, the more elegant.
>
> Ptolemy also tells us indirectly what was the attitude of his contemporaries toward theory and observation. It is clear that he expected his theories to be judged on the basis of their agreement with observation. If not, why did he fabricate so many data? Why did he not simply present his theories and let them stand or fall on the basis of their "rigor, logic and consistency"?[27]

I think that we have now considered enough cases. We have looked at several ancient Greek astronomers who come readily to mind, and we did not find a single one to whom "rigor, logic and consistency" were more important than "observational accuracy". In fact, we have every reason to believe that their attitude toward the proper relation between theory and observation was essentially the same as ours.

In fact, we find this attitude explicitly expressed by several Greek writers who say that the purpose of theory in astronomy is to "save the phenomena", in the wording often used in the modern writing on the subject. That is, a theory should meet two conditions: It should provide a means of making quantitative calculations, and the calculations should "save" or preserve or agree with the observed phenomena. Is this not what we say about the function of theory today?

Thus, in this regard at least, we may judge Ptolemy by the same standards that we would apply to a modern scientist.

8. A Final Summary

While we are not able to find specific documentary sources for many of the parameters that appear in Ptolemy's *Syntaxis*, there is no conceptual problem in understanding their origins. There is only one exception to this statement. Ptolemy takes the apparent diameter of the sun to equal the minimum diameter of the moon rather than its average diameter, in contradiction to the results of earlier astronomers and in contradiction to the known existence of annular eclipses. I suggested a

[27] I thank my colleague B. B. Holland for suggesting this argument.

possible explanation of this choice in Section IV.6, but it is only a speculation that I see no way to test.

In *CCP* I concluded not only that Ptolemy was an intellectual fraud but that there were serious reasons to doubt his general competence. The current study has revealed some errors that I overlooked in *CCP* and thus it strengthens our doubts. For example, Ptolemy made an error in calculating the length of the longest day on the tropics. He does not seem to understand the effect of rounding on the precision of his results. As I showed in Section VI.2, he rounds basic data used to find the mean motions of the planets in the fourth sexagesimal position and then proceeds to calculate the mean motion to six positions.

So far as I can see, Ptolemy almost never checks his work and, when he makes a mistake, he almost never corrects it, but goes on to fabricate data to prove that his error is in fact correct. An example concerns the apogee of Mercury, where his theoretical error causes him to get a value that is wrong by almost 30°, but which he then confirms by three sets of fabricated data. There may be one exception. I believe (Section VII.6) that Ptolemy originally intended to use the equant model for Mercury as well as for the other planets. Then, when he found that perigee was not 180° away from his erroneous position of apogee, he did not fabricate data to put perigee where he first expected. Instead, he replaced the equant model by his complicated model that does keep perigee from being opposite apogee. It is ironic that when he did this, he did not correct an error. He created still another one.

Even this example may not be a true exception. When Ptolemy introduced his complicated model of the motion of Mercury, the motivation may not really have been the correction of an error. It may have been an opportunity to display his virtuosity in devising models of planetary motion.

Ptolemy is also self-contradictory in the matter of the parallaxes of the inner planets. He finds (Section IV.6, above) that the horizontal parallax of the sun is almost 3 minutes of arc. He also concludes [*Syntaxis*, Chapter IX.1] that Mercury and Venus are closer than the sun, and in *Planetary Hypotheses* he even has the minimum distance of Mercury equal the maximum distance of the moon, which has a large parallax.[28] Yet Ptolemy says explicitly [Chapter IX.1] that none of the planets has a sensible parallax. Thus Ptolemy contributes nothing to the subject of parallaxes. Earlier astronomers had measured the parallaxes of the sun and moon, and Ptolemy confirms their results by fabricated data. Where earlier astronomers were silent on parallax, Ptolemy can do nothing new.

[28] It is possible that Ptolemy had not yet had this idea when he wrote the *Syntaxis*.

One peculiarity of the *Syntaxis* is that its tables of solar, lunar, and planetary positions all stop at a time that is about 60 or 80 years before the time of Ptolemy, as I remarked in Section VI.2. Now Ptolemy presumably wrote the *Syntaxis* for the use of himself, his contemporaries, and his successors. If this is so, it does not seem plausible to me that he would prepare his tables only for earlier times, without bringing them down even to his own time. This leads to the conclusion that his tables of position, like the star catalogue, were simply appropriated without acknowledgment from an earlier work.

Another peculiarity is that the table of latitude as a function of the longest day of the year was not calculated in Ptolemy's time nor was it calculated with the aid of his table of chords. It was calculated by someone, probably in the time of Eratosthenes and quite possibly by him, whose table of the chord function had only two sexagesimal positions instead of the three that appear in Ptolemy's table. See Appendix D for a further discussion of this point.

Ptolemy is often praised for the elegance of his mathematical methods and for his ability to choose observations in a way that simplifies the analysis. We see now that this is not quite so. In finding the maximum equation of the center at the quarter moons, Ptolemy needs to separate the problem of finding the longitude of the moon from the problem of finding its parallax. Every lunar day the moon passes through a position where the parallax in longitude is zero, and this position can be calculated without knowing the magnitude of the parallax. Ptolemy could have separated the parallax and the longitude by using observations made at this position, but he did not do so. Instead, he simply asserted incorrectly (see Section III.6) that the measurements were made under conditions of zero parallax when in fact it was fairly large.

For the sun, moon, and the outer planets, there is a point in the analysis where Ptolemy has to derive three parameters from three observations. He solves the problem for the sun by using what I call the method of the seasons, which had already been used by Hipparchus. This is a simple and elegant method, which depends for its simplicity upon choosing the observations in a particular way. As I shall show in Appendix B, Ptolemy could have chosen his observations of the moon and the outer planets this way, but he did not do so. Instead, he chose them in a way that forced him to a complicated and cumbersome method of solution.

I want to word this point carefully. It is often hard to think of an easy way to solve a problem, and we should not criticize Ptolemy for failing to apply the method of the seasons to the moon and outer planets.[29] However, we should not praise him for his ability to simplify the problem when in fact he failed to do so in every relevant case but one.

[29]After all, Hipparchus had also failed to do so, at least for the moon.

We have also seen (Section II.5) that Ptolemy probably chose to have the orbit of the sun precess along with the equinoxes in order to avoid two sets of calculations, but that this consideration did not apply with the planets. For them, he chose to have the orbits remain fixed with respect to the stars, and doing so did not complicate his analysis. Here, and I suspect in some other places, Ptolemy used what I euphemistically call the principle of economy of effort.

With regard to Mercury, we have found a plausible reason why Ptolemy thought that it had two positions of perigee in each revolution around its deferent curve, which is the point that led him to invent his complicated model of Mercury. This discovery, along with Ptolemy's reason for making the solar orbit precess with the equinoxes, and the application of the method of the seasons to the moon and planets, are perhaps the most interesting new results of the present work.

Until I wrote the present work, I had not realized how important astrology was in the work of Ptolemy. His *Tetrabiblos* is a manual of astrology, and his *Handy Tables* was almost surely devised for the use of astrologers. In fact, I suspect that astrology supplied much of the motivation for the *Syntaxis* and for much of ancient astronomy. However, I do not think that astrology supplied all of Ptolemy's motivation for working in astronomy. The cosmological scheme in *Planetary Hypotheses* does not have anything to do with astrology, as I understand it, since it deals with the distances of things while astrology is concerned only with their angular positions. Here, I think, Ptolemy wanted to be known as the man who built up a theory of the structure of the entire universe, and his motivation in at least this matter was astronomical rather than astrological.

I close this work with the same words I used to close *CCP*: "The *Syntaxis* has done more damage to astronomy than any other work ever written, and astronomy would be better off if it had never existed. Thus Ptolemy is not the greatest astronomer of antiquity, but he is something still more unusual: He is the most successful fraud in the history of science."

APPENDIX A

PTOLEMY'S USE OF THE WORD "CALCULATE"

Several times in the *Syntaxis* Ptolemy uses a word that is usually translated as "calculate" or "compute" in English and as *rechnen* in German. For example, in referring to the summer solstice of the year 140 in Chapter III.1 of the *Syntaxis*, he writes that "we very carefully calculated it to have fallen" at two hours after midnight on the day we call 140 June 25. Since he did indeed fabricate the time of this "observation", it at first looks as if Ptolemy unconsciously slipped and admitted that he found the time by calculation rather than by observation.

This is not the case, however, because Ptolemy does not use the word with the meaning that we now attach to "calculate" or "compute". In order to establish his meaning, we must analyze his usage, and the place in the *Syntaxis* where his meaning is most obvious is probably in Chapter IV.6

In this chapter, Ptolemy first introduces records of three lunar eclipses that were "observed" in Babylon in −720 and −719.[1] He next uses these records to find three parameters of the moon's orbit, as I describe in Section III.4 of the present work. He then "verifies" these results by using "three eclipses" that were "most carefully observed by us in Alexandria,. . ." The first occurred during the night between 133 May 6 and 7, and he later says of it: ". . . we accurately calculated the middle of it to have occurred ¾ equatorial hour before midnight." Similarly he says of the other two eclipses that he calculated their middle hours, which is the property of the eclipse that he needs in order to find the parameters under discussion.

[1] I put "observed" in quotation marks here because there is strong evidence that Ptolemy fabricated at least some of the data relating to these eclipses. It is quite possible, however, that there were genuine records of these eclipses still available in his time, and that Ptolemy altered the data found in them. See *CCP*, Section VI.6.

In the later passages he says that he "calculated" the times even though he explicitly says earlier that he "observed" the eclipses. Thus, at least in referring to observations, he uses "observe" and "calculate" almost interchangeably. The reason is not hard to find. In the case of the eclipses, for example, it is difficult and inaccurate to judge when the magnitude of the eclipse is a maximum, and the middle hour, if found directly by observation, would be rather inaccurate. It is far more accurate to measure the times of the beginning and end of the eclipse, and to find the middle hour by calculation from these measured times.

Similar argument applies to the summer solstice. Ptolemy could not have found the time by direct observation, because he could not have seen the sun at two hours after midnight. The way to "observe" the time of a summer solstice was and is to measure or observe the meridian elevation of the sun on a suitably chosen series of days, and to find the time of the solstice by calculation from these values. See Section II.9.

Thus, in a context such as this, Ptolemy uses the word that is translated as "calculate" to indicate what we often call "reduction of data". By this, we mean the set of calculations that takes us from "raw data" to "data" in the form needed for theoretical analysis. However, to speak of reducing data automatically implies that there are observed data to be reduced. Thus Ptolemy implies that he observed the summer solstice of 140, even though he writes that he calculated its time. With regard to the lunar eclipses, Ptolemy explicitly says both that he observed them and that he then used calculation to give him the middle hours that he needed. (At the risk of being superfluous, I point out that the hours were in fact fabricated.)

In my writing on the *Syntaxis*, I have not tried to distinguish between the occasions when Ptolemy says that he calculated some observational quantity and the occasions when he says that he observed it, because his use of "calculate" automatically implies the existence of previous observations in these contexts. I use "observe" on all these occasions because I have not thought of a simple translation that preserves his meaning accurately.

APPENDIX B

THE APPLICATION OF THE METHOD OF THE SEASONS TO THE MOON AND THE OUTER PLANETS

1. The Application of the Method to Jupiter

In Section II.5, I described the "method of the seasons" that Ptolemy, and Hipparchus before him, used to find the apogee a, the eccentric e, and the anomaly γ at some particular epoch for the sun. In this method, which seems rather elegant to me, one first finds a and e simultaneously and then finds γ in a separate calculation.

In Sections III.5 and VIII.4, I pointed out that this method can be applied without change to the corresponding problems for the moon and for the outer planets.[1] In this appendix, I shall present a test of this method for the moon and for Jupiter, which I continue to take as an example of an outer planet; since Ptolemy's treatment is identical for the three outer planets, we need to study only one of them in detail. I shall take up Jupiter first and then turn to the moon in the next section.

Since Ptolemy does not present the data that we need in order to test the method, our first job is to "fabricate" the needed data. However, in order to see that the method is applicable to Ptolemy's problem, we should use the same kind of data that he did. These data are measurements of the time when Jupiter passed through mean opposition, in which the true position of the planet is opposite to the mean position of the sun.

Some years ago, for a different purpose, I programmed Hill's theory of Jupiter [*Hill*, 1895] and Newcomb's theory of the sun

[1] The problem does not arise with the inner planets, where the main problem is the study of the planets at their maximum elongation from the sun. The entire methodology is different for them.

[*Newcomb*, 1895]. By combining these programs, I have calculated the circumstances of seven mean oppositions that occurred between 129 and 135. In the calculations, I first found the mean longitude L_S of the sun and the geocentric longitude λ of Jupiter at intervals of one day near an opposition, and I then calculated the elongation D at these times using $D = \lambda - L_S$. I then found the time of the mean opposition and the longitude of Jupiter at that time by interpolating to make D equal to 180°, using linear interpolation.

Table B.1 contains the needed circumstances of the seven mean oppositions. The first column gives the date of the mean opposition, the second column gives the hour, in mean time at Alexandria, and the third column gives the geocentric longitude of Jupiter at this hour. Ptolemy claims that he measured the mean opposition of 133 May 17, that he found it to come at 23 hours, and that the longitude of Jupiter was 233;11 degrees. However, we remember that his longitudes are referred to a wrong position for the equinox, so we should add 1°.1 to his longitude, getting 234°.183. Table B.1 gives 22.49 hours for the time and 234°.154 for the longitude. However, Ptolemy's claimed observation is not as good as this comparison suggests. In the calculations that lead to Table B.1, I have neglected the secular deceleration of the earth's rotation that is produced by tidal friction and other effects. This means that the times in Table B.1 are biased by an amount that is probably of the order of 3 hours, so Ptolemy's time is also in error by approximately this amount.

TABLE B.1

SOME MEAN OPPOSITIONS OF JUPITER[a]

Date	Hour, Alexandria Mean Time	Geocentric Longitude of Jupiter, *degrees*	Change in Mean Longitude of Jupiter,[b] *degrees*
129 Jan 13	9.91	111.386	0.000
130 Feb 13	17.35	142.009	32.930
131 Mar 16	12.50	172.126	65.817
132 Apr 15	15.79	202.577	98.733
133 May 17	22.49	234.154	131.743
134 Jun 20	22.92	267.445	164.898
135 Jul 26	20.42	302.587	198.209

[a]Calculated from the theories of *Hill* [1895] and *Newcomb* [1895].
[b]Calculated using 0.083 091 22 degrees per day for the mean motion.

TABLE B.2

LONGITUDE OF THE EPICYCLE CENTER OF JUPITER AS A FUNCTION OF THE CHANGE IN ITS MEAN LONGITUDE

Longitude of the Epicycle Center, *degrees*	Change in Mean Longitude of the Epicycle Center Since the Mean Opposition of 129 January 13, *degrees*
111.386	0.000
201.386	97.445
291.386	187.592

The last column gives the change in the mean longitude L of Jupiter that occurs after the time of the first mean opposition. I calculated these changes by using $L' = 0.083\ 091\ 22$ degrees per day.[2] This is taken from Hill's tables and reduced to the epoch 1960 January 1 [*Explanatory Supplement*, 1961], so it is probably not very accurate for Ptolemy's time. However, we do not need an accurate value over the short time span of Table B.1. Using this provisional value of L' is analogous to Ptolemy's using a provisional value of γ' in his solution for the elements of Jupiter, even though he had not yet found it from observations.

The last two columns of Table B.1 give the longitude λ of the center of Jupiter's epicycle[3] as a function of the change ΔL in its mean longitude since the opposition of 129 January 13. The table gives this function for somewhat irregular intervals in both variables, but we want it for uniform intervals of 90° in the longitude λ of the epicycle center. We find the change ΔL in the mean longitude for these values of λ by interpolating linearly between the oppositions of 131 March 16 and 132 April 15 for the first value and between those of 134 June 20 and 135 July 26 for the second value.

The resulting values are given in Table B.2, and with them we are now ready to apply the method of the seasons. We do so with the aid of Figure B.1. In order to draw the figure correctly, we must first find the quadrant in which apogee lies. We see from Table B.2 that during the first "season" the mean longitude changes by 97°.445, while it changes by only 90°.147 during the second "season." Thus apogee lies between the first two values of λ in Table B.2. Not knowing the exact value of apogee yet, I have drawn the apogee midway between the first two values of λ. Otherwise I have drawn the angles in Figure B.1 to scale.

[2]This is 299.1284 seconds per day.

[3]The reader should remember from Section VIII.3 that the longitude of the planet is the same as the longitude of the center of its epicycle at the time of a mean opposition.

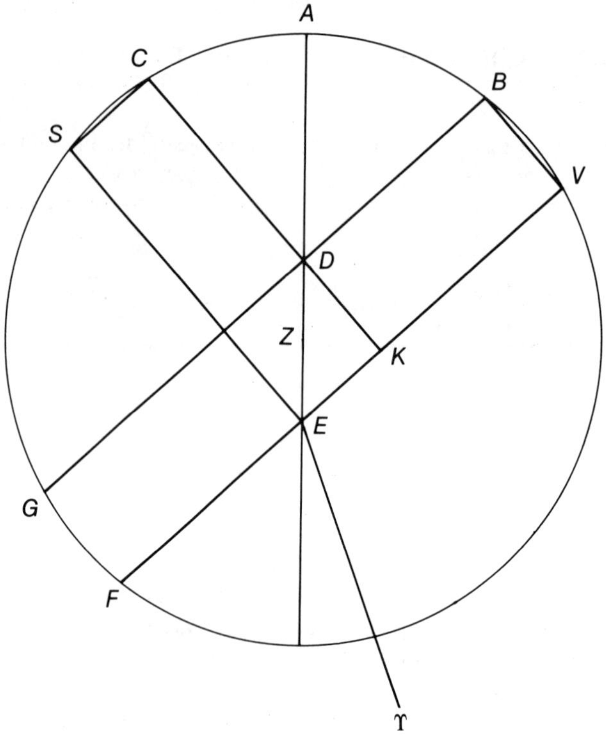

Fig. B.1—The method of the seasons as applied to an outer planet. *E* is the earth. Points *V*, *S*, and *F* are three positions of the planet when it is at mean opposition, with the longitudes being separated by multiples of 90°. Two of the three points may be determined by interpolation between observations, if necessary, but one point should be the result of a single observation. The solution is exactly like that for the sun in Figure II.4.

The lettering in Figure B.1 is the same as in its solar counterpart Figure II.4 except for one change: Ptolemy uses the simple eccentric model for the sun, so that the mean sun moves uniformly around *Z*, which is the center of the circle. For Jupiter, however, Ptolemy uses the equant model, in which the center of uniform rotation is the point *D* rather than the center of the circle.

The center of the epicycle is viewed from the earth *E*, and points *V*, *S*, and *F* are the three positions of the center listed in the first column of

Table B.2. Lines parallel to *VEF* and *ES* are drawn through point *D*, which is the center of uniform rotation, rather than through the center *Z* of the circle. Following Ptolemy, I take the distances *EZ* and *ZD* to be equal, and I represent each of them by *e*.

The solution for *a* and *e* goes just as it did in Section II.5, with one small change. Arc *VSF* is 187°.592 so arc *BV* is 3°.796. Instead of using the sine of this angle, we use its chord, so that the line *DK* is 0.066 241. Arc *SC* is 3°.649, so the line *EK* is 0.063 676. Line *DE* is then 0.091 884, but this is 2*e*, so *e* equals 0.045 942, and the angle *DEK* is 46°.131. Since the longitude of *V* is 111°.386, apogee is at 157°.517.

To find the anomaly at the epoch of the mean opposition of 129 January 13, which corresponds to point *V* in Figure B.1, we go to Figure B.2. I have now left out all points in Figure B.1 except point *V* and the points along the line *EZDA*. To find the anomaly, we must find the equation of the center, which is the sum of the angles δ_1 and δ_2. The angle β is 46°.131, as we just found, the two distances marked *e* are also known, and we are taking the radius *ZV* of the circle to be 1. When we

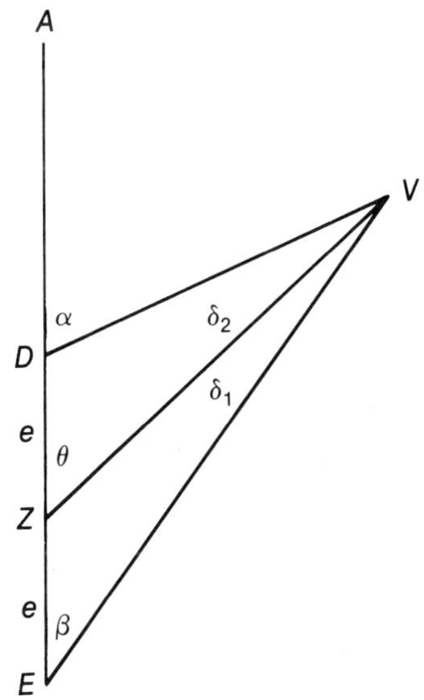

Fig. B.2—Finding the equation of the center in the equant model from an observed longitude. The angle β is known from observation, and the equation of the center is the sum of the angles δ_1 and δ_2. The equal distances *e* and the length of the line *ZV* are known from the model.

apply the law of sines to triangle ZEV, we immediately find $\delta_1 = 1°.898$. Thus the angle θ equals 48°.029.

When we apply the law of sines to the triangle DVZ, we get

$$\sin \delta_2 = e \sin (\theta + \delta_2),$$

which must be solved by successive approximations. We know that the value of δ_1 is a good approximation to δ_2. If we start with it, the second approximate solution is 2°.018, which is correct to the three decimal places stated. Thus the equation of the center is 3°.916.

The longitude of point V is 111°.386 from Table B.2, so the longitude L_S of the mean sun is 291°.386. It is clear from the figure that the true longitude of V is greater than its mean longitude L, so that we have $L = 111°.386 - 3°.916 = 107°.940$. By Equation VIII.6, $L_S = L + \gamma$, in which γ is the anomaly, so $\gamma = 183°.916$.

Instead of using the era of Nabonassar as the epoch to which γ_0 is referred, I should like to use noon, Alexandria time, on 137 July 20,[4] which is 3 110^d.091 after the time of the observation. The mean motion of the sun is 0.985 647 34 degrees per day and the mean motion of Jupiter is 0.083 091 22 degrees per day. Hence the daily change in the anomaly is 0.902 556 12 degrees, and the total change in anomaly between the observation and the reference epoch is 287°.032, exclusive of the complete circles. Thus the anomaly at the epoch is 110°.948.

In Table XI.5 of *CCP*, I give the parameters that best fit Jupiter's orbit, as calculated from modern theory. However, in making that fit, I allowed the two eccentrics to be different, and I have never made a fit in which I forced them to be the same. Thus the comparison of those parameters with the ones found here is not particularly significant. Accordingly, the only comparison I can make here is with Ptolemy's values. This is done in Table B.3.

The only appreciable difference between the parameters is in the position of apogee; the two values differ by about 4°.5.[5] This makes a maximum difference of about 0°.2 in a calculated position.

The solution found here should not agree with the best-fitting solution, if we knew what it was, for two reasons: (a) the solution found here is based upon only a few positions instead of upon a great number, and (b) I used linear interpolation in Table B.1 over a rather large interval of about 35°. It would probably have been better to use mean conjunctions as well as mean oppositions. This would have halved the interval and divided the error in interpolation by a factor of 4.

[4] I choose this epoch because it is the epoch that Ptolemy chooses for the star catalogue in the *Syntaxis*. It is the first day of an Egyptian year.

[5] In preparing Table B.3, I have neglected the small change in apogee between the observation of 129 January 13 and the epoch of 137 July 20.

TABLE B.3

SOME SOLUTIONS FOR CERTAIN PARAMETERS OF JUPITER

Parameter	Method of the Seasons	Ptolemy's Method
Apogee at the epoch,[a] *degrees*	157.517	162.1[b]
Eccentric[c]	0.045 942	0.045
Anomaly at the epoch,[a] *degrees*	110.948	110.936

[a]Noon, Alexandria mean time, on 137 July 20, rather than the era of Nabonassar.

[b]Ptolemy actually obtains the value 161°.0, but this is referred to the wrong position of the equinox. I have added 1°.1 in order to refer his solution to the correct equinox.

[c]Referred to unit distance rather than to 60.

We should note that Ptolemy also uses interpolation on occasion. He establishes the relation between the maximum elongation of Mercury and the position of the mean sun (Section VII.2), and he then interpolates between positions of the sun to give him the values of the maximum elongation that he needs.

2. The Application to the Moon

Since Ptolemy uses lunar eclipses in finding the parameters of the moon, we should do the same thing. The first eclipse he uses is that of −720 March 19, observed in Babylon, and I decided to make this the starting point of the test. In order to get some "data", I used Oppolzer's canon of eclipses [*Oppolzer*, 1887] to find the next 19 eclipses visible in Babylon after this one, for a total of 20. From these 20, I selected five in which the circumstances were suitable for the test.

In order to use the method of the seasons, we need the position of the moon at the center times of the eclipses. Since Oppolzer does not tabulate the position, it is necessary to calculate it. I have computer programs to calculate the lunar position from modern theory but, for the sake of variety, I decided to use Ptolemy's theories of the sun and moon instead.

The first task in doing this is to find the times of the eclipses according to Ptolemy's theory. To do this, I first calculated the positions of the sun and moon from his theory using Oppolzer's times, and subtracted the solar longitude from the lunar one to get the true elongation. This differed from 180° by a maximum of 0°.482. Using the average rate at which the elongation changes, I then estimated a new eclipse time, and I calculated the longitude of the moon at this time. I did not make a new calculation of the elongation, feeling that the time and position found this way are accurate enough.

As a matter of amusement, since the eclipse of −720 March 10 was observed at Babylon, I transferred the calculated times to that place, using Ptolemy's value (50 minutes) for the time difference between the two places.

The characteristics of the eclipses that we need are listed in Table B.4. The first column gives the date and the second column gives the calculated hour at Babylon. The third column gives the longitude λ of the moon at the time when the true elongation was estimated to be 180°, by the method just described. The next two columns give the changes in apogee a and anomaly γ between the time of the first eclipse and the time of any of the later ones, using Ptolemy's values for the ratios of change of a and γ. The last column needs some explanation.

TABLE B.4

SOME LUNAR ECLIPSES VISIBLE AT BABYLON[a]

Date	Hour at Babylon	λ, *degrees*	Δa, *degrees*	$\Delta\gamma$, *degrees*	$\lambda - \Delta a$, *degrees*
−720 Mar 19	21.74	174.642	0.000	0.000	174.642
−719 Sep 1	20.32	333.323	59.147	96.735	274.176
−713 Oct 24	20.92	25.980	309.130	254.883	76.850
−710 Feb 27	20.11	154.536	44.595	291.129	109.941
−710 Aug 23	20.15	324.286	64.313	83.657	259.973

[a]Calculated from the lunar and solar theories in the *Syntaxis*.

Ptolemy postulates that the longitude of apogee is constant for the sun and that it changes by 1° per century for the planets. Since the data in Table B.1 cover a span of only 6 years, we may neglect the change in apogee and take it to be constant for the planets also. For the moon, however, apogee advances rapidly and makes a complete revolution in less than 9 years. That is, apogee makes more than a full revolution during the time covered by the data in Table B.4.

Thus, in order to apply the method of the seasons to the moon, we must refer the longitude to a direction that rotates at the same rate as apogee.[6] We accomplish this by subtracting the change Δa listed in Table B.4 from the longitude of the moon. This gives us the last column in the table, which is labelled $\lambda - \Delta a$. This column will play the role played by the longitude in the applications to the sun and the outer planets, while $\Delta\gamma$ will play the role formerly played by the mean longitude.

By means of the data in Table B.4, we can interpolate to find values of $\Delta\gamma$ when $\lambda - \Delta a$ takes on the values 174°.642 ± 90; 174°.642 is the value at the time of the first eclipse in the table. This gives us the values in Table B.5. Since the first value of $\Delta\gamma$ is 263°.418, I have let the other values equal or exceed 360° in order to avoid a discontinuity.

TABLE B.5

SOME QUANTITIES USED IN THE THEORY OF THE MOON

$\Delta\gamma$[a], *degrees*	$\lambda - \Delta a$[a], *degrees*
263.418	84.642
360.000	174.642
447.956	264.642

[a] $\Delta\gamma$ and Δa are referred to the relevant values at the middle of the eclipse of −720 March 19.

Ptolemy's model for the motion of the moon is quite complicated, but it reduces to a simple epicycle at the syzygies. Since the epicycle and eccentric models are equivalent, I replace the epicycle momentarily by the equivalent eccentric in order that we may use the same figure as before (Figure B.3). The first "season" (increment in $\Delta\gamma$) in Table B.5 gives a change of 96°.582 while the second one gives a change of only 87°.956. Thus the direction *EA* should again lie in the first season.

[6] We cannot refer the longitude to the apogee itself, because the position of apogee at a reference epoch is one of the parameters to be found.

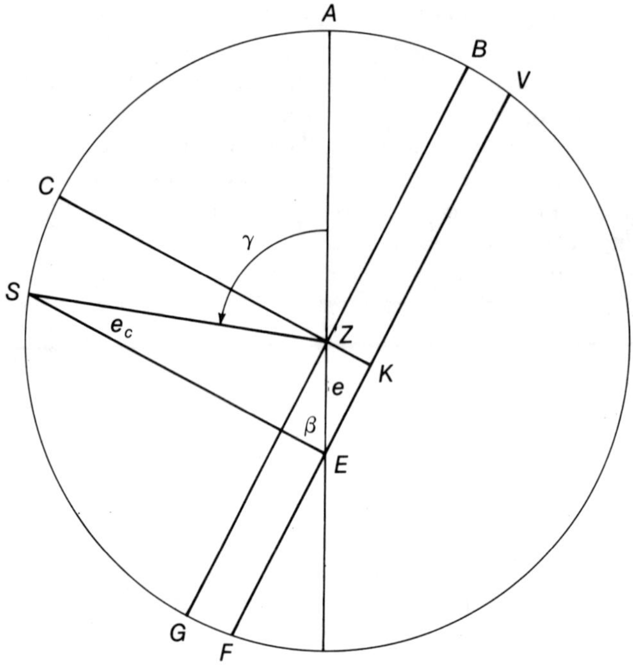

Fig. B.3—The method of the seasons as applied to the full moon. The moon is viewed from the earth *E*. The points *V*, *S*, and *F* are not actual positions as they were for the corresponding Figures II.4 and B.1. Instead, they are longitudes of the full moon minus the value of apogee, except for a constant bias that is to be found from the data. The line *ES* is perpendicular to the line *VEF*. Two of the three longitudes may be found by interpolation, but at least one must be the result of a single observation. The solution of the figure is almost exactly like that for the sun in Figure II.4 and for an outer planet in Figure B.1.

The values of Δa in Table B.4 are referred to the time of the eclipse of −720 March 19, when the longitude of the moon was 174°.642, and this is the only actual position that appears in Table B.5. Thus the direction *EA* is the direction of apogee at the time of this eclipse.

The solution goes exactly like the solution for the sun in Section II.5, since we are again dealing with the simple eccentric model, and I shall only state the results:

$$e = 5.099, \qquad \text{angle } ZEK = 27°.765. \qquad \text{(B.1)}$$

In this, I have gone back to using 60 for the radius of the circle rather than unity. The value of e becomes the radius r of the epicycle when we go back to the epicycle model.

Since angle ZEK is 27°.765, the angle β in the figure is 62°.235. Since the longitude of the point S as seen from E is 174°.642, the longitude of A is 112°.407. That is, a is 112°.407 at the time of the eclipse of −720 March 19. It remains to find the anomaly γ at this epoch.

In Figure B.3 I have drawn in the line ZS, which does not appear in the corresponding Figures II.4 and B.1. The object now is to solve the triangle SZE, in which the sides SZ (equal to 60) and ZE (equal to the eccentric distance 5.099) and the angle β are known. By using the law of sines, we immediately find that angle e_c equals 4°.313. γ is clearly the sum of β and e_c, so it equals 66°.548.

We now have the complete solution to the problem. I have also found the solution that best fits positions of the moon calculated from modern theory, but it does not apply here because here I have "fabricated" the data using Ptolemy's theory. Since this is so, however, the solution found here should agree accurately with Ptolemy's solution.

Table B.6 gives the comparison of this solution with Ptolemy's solution. The most important difference is probably that between the two values of the epicycle radius. This is about 0.15 parts out of 60, which can make a maximum change of about 8′ in a calculated longitude. This is far less than the errors in Ptolemy's theory, so the method of the seasons is just as valid as his method.

TABLE B.6

A COMPARISON OF TWO SOLUTIONS
FOR THE LUNAR PARAMETERS

Parameter	Present Solution	Ptolemy's Solution
Apogee at the epoch,[a] *degrees*	112.407[b]	112.890[b]
Radius of the Epicycle[c]	5.099	5.250
Anomaly at the Epoch,[a] *degrees*	66.548	66.171

[a] The epoch is the middle of the eclipse of −720 March 19.
[b] These values are referred to Ptolemy's erroneous position of the equinox.
[c] Referred to a deferent radius of 60.

The discrepancy between the two solutions probably arises from the need to interpolate in Table B.4. In getting the value 264°.642 for $\lambda - \Delta a$ in Table B.5, I interpolated between the values for −719 September 1 and −710 August 23. These values are separated by only about 14° and this interpolation is probably rather accurate. In getting the value 84°.642, however, I had to interpolate in an interval of about 33°, and this interpolation may have introduced error.

As I have said before, I believe that Ptolemy is wrong in insisting that the theory at the syzygies should be based upon lunar eclipses. I showed in *CCP* [p. 123] that the standard deviation in a longitude found by the eclipse method is about 15′, which is about the same as the error in a direct measurement of the longitude. If Ptolemy had used measurements of the longitude of the moon when it is approximately full, he would have had a dense set of data and he could have used the method of the seasons conveniently and accurately.

One other feature of Ptolemy's method should be pointed out. Ptolemy always neglects the evection term in the position of the moon at the center of an eclipse. However, at this time, it is the true and not the mean elongation that is 180°. The evection depends upon the mean elongation, which can easily range from 175° to 185° or more at the center of an eclipse. The evection term can amount to as much as about 9′.

In summary, both for the moon and Jupiter, the biggest error in applying the method of the seasons probably arises from the need to interpolate between observations. Ptolemy could easily have cut down on the size of the interpolation error by using additional types of data in order to have observations at small intervals.

APPENDIX C

PTOLEMY'S TABLE OF THE CHORD FUNCTION

Ptolemy gives a table of the chord function in Chapter I.11 of the *Syntaxis*, and he gives a table of the latitude as a function of the longest day of the year in Chapter II.6. I shall study the table of latitude versus the longest day in Appendix D. I shall study the table of chords in this appendix, both because it is interesting in its own right and because it is needed for Appendix D.

The chord function was defined by Equation I.1 in Section I.4, which I repeat here for convenience:

$$\text{chord}\, x \;=\; 2 \sin \tfrac{1}{2} x \;. \qquad \text{(C.1)}$$

I shall generally abbreviate "chord x" as "chd x" in this appendix and in Appendix D. A modern mathematician, in tabulating the chord function, would probably give the length of the chord inscribed in a circle whose radius is unity. Thus he would write, for example, chd 60° = 1 and chd 180° = 2. Ptolemy, however, gives the length of the chord inscribed in a circle whose radius is 60. Thus, if he were to use the equals sign, he would write chd 60° = 60 and chd 180° = 120.

Ptolemy also uses the sexagesimal notation rather than the decimal notation, and he uses three sexagesimal positions in giving the chord function. Thus, in full, he writes chd 60° = 60;0,0. This gives about twice the precision that is provided by five decimal positions.

Ptolemy gives the value of the chord for every half degree from ½ degree to 180 degrees (he omits the entry for 0°). Thus there are 360 lines in his table.

In Chapter I.10 of the *Syntaxis* Ptolemy explains how he found the chord for every multiple of ½ degree; I summarized the method in Section II.3 of *CCP*. To start with, we know chd 180°, chd 90°, and chd 60° from the properties of elementary triangles. In addition, the regular pentagon and the regular decagon had been completely solved in Euclid and it was known that the ratio of the side to the radius of the circumscribed circle is 1.175 570 505 for the pentagon and 0.618 033 989 for the decagon.[1] This means that chd 72° = 70;32,3 and that chd 36° = 37;4,55, in Ptolemy's system.

Next, if we are given chd x, it is almost trivial to find chd ½x; the relation between chd x and chd ½x is the analogue of the relation $\sin x = 2 \sin \frac{1}{2}x \cos \frac{1}{2}x$. The application of this relation to the values already known gives the chords of all angles of the form $90°/2^n$, $72°/2^n$, and $60°/2^n$, in which n is an integer. It also gives chd 120° and chd 144°.

In order to proceed further, Ptolemy uses a theorem which is equivalent to the theorem that lets us calculate $\sin (x \pm y)$ from $\sin x$ and $\sin y$. By applying this theorem to the values already obtained, Ptolemy can find the chords of all multiples of 1½ degrees, and of all angles that can be obtained by successive halving of 1½ degrees,[2] but he cannot yet find chd ½, which is the final thing he needs to prepare his table.

In order to find chd ½, Ptolemy first finds chd 1°, and in order to find that, he establishes the theorem that (chd x/chd y) $< x/y$ when $x/y > 1$. He then uses two special cases of this theorem:

$$\text{chd } 1° < (4/3) \text{ chd } \tfrac{3}{4} \quad \text{and} \quad \text{chd } 1° > \tfrac{2}{3} \text{ chd } 1\tfrac{1}{2} \ .$$

From these, he finds

$$\text{chd } 1° < 1;2,50 \quad \text{and} \quad \text{chd } 1° > 1;2,50 \ , \qquad \text{(C.2)}$$

simultaneously, to the accuracy given. Hence to three positions, chd 1° = 1;2,50. I shall repeat two comments that I made in Section II.3 of *CCP*.

First, while chd 1° does equal 1;2,50 to the accuracy given, Ptolemy does not prove it. The first of the relations in C.2 is not correct to the accuracy given; it should read chd 1° < 1;2,51. Thus Ptolemy cannot tell from his method whether to take 1;2,50 or 1;2,51 for chd 1°. He may simply be lucky that he chose the right value. Alternatively, the value 1;2,50 may have been established by a correct method in earlier work that is now lost, and Ptolemy may have mistakenly thought that he had found a simple proof of it.

Second, Ptolemy says that "the chord of a third of the arc is in no way geometrically given, . . ." *Pedersen* [1974, pp. 60 – 61] takes this to mean that finding chd ½ from chd 1½ " . . .would imply the trisection of an angle; but that was one of the problems which Greek

[1]I do not know the number of significant figures to which these values had been carried in Ptolemy's time.

[2]On page 26 of *CCP* I inadvertently wrote that one could find the chords of all integral fractions of 1½ degrees. I meant all fractions of the form $1\frac{1}{2}/2^n$.

mathematics had never succeeded in solving by methods involving construction with ruler and compasses only." (As he also points out, we know now that trisection of an angle is impossible with ruler and compass only.)

I agree that Ptolemy was probably referring to trisection in his remark. If so, the matter does not deserve the emphasis he gives it, because there are at least two simple ways to find chd ½ degree from chd 1½ degrees, even though we cannot trisect an angle by ruler and compass. Finding chd ½ is the only thing that matters in the context of the remark.

In the first method available to Ptolemy, he could find chd $2x$ from chd x and chd $(x + y)$ from chd x and chd y. Thus he could easily relate chd $3x$ to chd x. In fact (*CCP*, p. 28),[3]

$$\text{chd}\, 3x = 3\, \text{chd}\, x - (\text{chd}\, x)^3 \,. \qquad \text{(C.3)}$$

The easiest way to solve Equation C.3 for chd x is probably by successive approximations, a method that was well known to Ptolemy. We start by taking chd x = ⅓ chd $3x$. Then we use the cube of this estimate to refine the estimate, and so on. In fact, chd 1½ = 1;34,15 in Ptolemy's table, and a third of this is exactly 0;31,25. This equals chd ½ to the accuracy given, and it is the value Ptolemy gives in his table. The use of relations C.2 is unnecessary.

In another simple way to find chd ½, we start from the inequality that leads to relations C.2; this inequality was well known long before Ptolemy. From it, we find

$$(\text{chd}\, 1\tfrac{1}{2})/(\text{chd}\, \tfrac{1}{2}) < 3 \,.$$

The question is the amount by which the ratio deviates from 3. Ptolemy could have taken the ratio to be 3 to find a starting estimate for chd ½. He could then have calculated chd 1½ carefully from this estimate, assuming it to be exact. By comparing the value of chd 1½ found this way with the value already known, he would have known whether the first estimate of chd ½ needed changing, and, if so, in which direction. In the actual case, no change would have been needed, to the accuracy of three sexagesimal positions.

By using a hand-held calculator that works to ten decimal digits, I have calculated the chord for every half-degree from ½ to 180°. I first calculated the chord in decimal notation and then converted the result to

[3]Note that chd $3x$ is a simple algebraic function of chd x. In fact, although the sine function (or the chord function) is a transcendental function, the values of the sine or chord that appear in ordinary trigonometric tables are algebraic numbers, not transcendental ones.

sexagesimal form. The comparison of these calculations with Ptolemy's table is given in Table C.1. In this table, I give only the entries for which my results differ from those of Ptolemy in the third sexagesimal position or earlier. The table appears at the end of this appendix.

The first line of the table, for example, refers to chd 2°. I find that chd 2° = 0;2,5,39.44, while Ptolemy gives 40 for the third sexagesimal position.[4] When the correct value is rounded to the third position, the value is 0:2,5,39 rather than 0;2,5,40. The actual error in Ptolemy's value is 0.56 of a unit in the third position.

Near the beginning of the table, the errors are random in sign, and occur only when the fractional part of the third position is fairly close to 0.5. In other words, the calculations back of Ptolemy's table were not carried to enough precision to give the correct rounding in critical cases. In the rest of the table, however, Ptolemy's value is almost always too large when it is in error, and the error exceeds a full unit of the third position in a number of cases.

Since chd 180° is known exactly from simple principles, I expected the incidence of error to go down as we come to the end of the table, but this does not happen. Two of the largest errors come for angles of 176° and 178½°. By analyzing the structure of the errors, it might be possible to determine exactly how Ptolemy's table was constructed, but I have made no attempt to do so.

In Appendix D we shall be interested both in the error of finding the chord when the angle is given and in the inverse error of finding the angle when the chord is given. I believe that the worst error in the inverse comes from using Ptolemy's value for chd 178½. The value he gives, namely 119;59,24 for a radius of 60, is actually the chord of 178°.5206. The error in finding the angle from the chord is thus 0°.0206. However, we are really interested in half the angle in the important cases, so the important error is only 0°.0103. This is about 37″.

It is not fair to take this as representative, however. In the first place, it corresponds to one of the largest errors in the table. In the second place, we should not try to find the angle from the chord when the angle is near 180°, just as we should not try to find the angle from the sine when the angle is near 90°. Instead, we should recast the problem in terms of the complement of the angle.

The next largest error is probably that made when the angle is 159½ degrees. Ptolemy gives the chord as 118;5,7, but this is actually the chord of 159°.5021. This corresponds to an error of 3″.8 in finding half the angle, and we are still at a place in the table where the chord is rather in-

[4]In this immediate discussion, I am taking the radius of the circle to be 1 rather than 60. I am giving the third sexagesimal position to a precision of a hundredth of a unit in this position. That is, I am using a mixture of sexagesimal and decimal arithmetic, which is common in astronomical literature. See Section I.4.

sensitive to the angle. Most errors will be smaller than this, but even an error this large is negligible in the considerations of Appendix D.

Pedersen [1974, p. 65] cites a study of the chord table made in 1900 that I have not consulted.[5] The conclusion of the study is that values of the sine derived from the table of chords are all accurate to the fifth decimal place. If by this is meant that the error is less than 0.00001, I believe that the statement is correct. Nonetheless, if we prepare tables of the sine from Ptolemy's table and from modern calculations, and if we round both tables to the fifth decimal place, we shall find numerous discrepancies in the fifth place. That is, Ptolemy's methods of calculation were not sufficiently accurate to give the correct rounding in a number of cases.

By now, the reader has probably been asking for a long time whether we really know the actual values of the chord that Ptolemy wrote down in his table. After all, many copyists have intervened between the original and the oldest copy that we still have. Is it possible that all the copies of such a large body of numerical values have been made without scribal error?

The answer is no. There are scribal errors in existing texts of the chord table. I have compared the Greek texts of Halma and Heiberg (see the entry for Ptolemy in the list of references), and I find that they differ in 21 values of the chord out of 360 values, which is a fairly large incidence of error. Further, in one of these, namely chd 88½, there are two independent differences, for a total of 22 out of 360. One difference was by 1 unit in the first sexagesimal position, one was by 9 units in the second position, and another was by 3 units in this position. All other differences were in the third position. One was by 30 units, 2 were by 3 units, 3 were by 2 units, and the remaining 13 were by 1 unit.

In all instances in which the difference is as large as 3 units in the third position, one text gives the mathematically correct value while the other does not. In these cases, I believe that the correct value must also be the original: It is unlikely that a copyist, in making a copying error, would accidentally alter the value from a wrong one to a correct one. It is more likely that the correct value is also the original one.

More accurately, we should say that the correct value was the one that Ptolemy intended to write. We may be sure that he knew the correct values of the chord, aside from uncertainties of one or two units in the third sexagesimal position. We may also be sure that he intended to write the values that he thought were correct, but that he made some scribal errors in doing so. We must also admit the possibility that some later copyist had a correct (and therefore independent) knowledge of chords, and that he (the copyist) substituted the correct value whenever he noticed an

[5]The reference is A. von Braunmühl, *Vorlesungen über Geschichte der Trigonometrie*, volume I, Leipzig, 1900.

erroneous value in the table from which he was copying. If so, the copyist departed from what Ptolemy actually wrote, but he restored what Ptolemy intended to write.

It seems almost certain that this has happened to some extent, as we can see from the distribution of the textual differences. Nineteen differences occur in the third sexagesimal position, and thirteen of these, more than half of the total, are differences of only one unit. This is highly unlikely if the differences resulted from scribal errors, which should occur equally in all positions. Thus it seems that copiers of the chord table have been guided to a considerable extent by the correct values, and that they did not introduce errors at random.

Where the difference is as small as one or two units in the third sexagesimal position, we cannot be sure what value Ptolemy intended. Thus copiers of the chord table could not be guided by the correct values in making such small corrections.

The most recent version of the chord table that I know of is the one given by Taliaferro in his English translation that is cited in the references. Whenever there is a difference between the texts of Halma and Heiberg, Taliaferro agrees with one text or the other in every instance but one, and he agrees with Halma about as often as with Heiberg. Further, whenever we can be sure of the value that Ptolemy intended to write (which is the correct value whenever the difference is more than 2 in the third place), Taliaferro's value is the correct one.

In addition, for chd 141½, which is the only instance in which Taliaferro's value disagrees with both Halma and Heiberg, Taliaferro's value is the correct one.[6] This suggests that Taliaferro based his version of the table upon the correct values, but we see immediately that this is not so: In many instances in which the difference between the texts is 1 or 2 in the third place, Taliaferro follows the text which gives the wrong value instead of the text that gives the correct one.

To sum up, Taliaferro has corrected errors that appear in the texts of both Halma and Heiberg. His version of the table is closer to the version that Ptolemy wrote, or at least intended to write, than any other version I know of, and it is probably as close as we shall ever be able to come to Ptolemy's intentions. If there are any remaining discrepancies between this version and Ptolemy's intended original, they are few in number compared with the errors in the original.

Thus I have based the study of errors in Table C.1 upon Taliaferro's version of the chord table.

In Table C.2 I have listed the angles for which Halma and Heiberg give different values of the chord, but I have not taken the space to list their values and the correct values. The interested reader should have no

[6]Here the difference between the Greek texts is 2 in the third place, and Taliaferro's value is the value between.

trouble in uncovering the actual differences in the texts with the aid of Table C.2. Table C.2 follows Table C.1 at the end of this appendix.

There has been considerable discussion about the originality of Ptolemy's table of chords. Now Greek astronomers as far back as Eratosthenes seem to have been making trigonometrical calculations with moderate accuracy (see Section II.7 and Appendix D), and it is likely that they used tables of some sort in the process. Thus the correct question to ask is whether Ptolemy developed a table that was more accurate than earlier tables, or whether he appropriated his chord table, like his star table, from an earlier source that he failed to acknowledge. I see no way to settle this question on the basis of available data.

Various writers have expressed admiration for the elegance of Ptolemy's treatment of the chord function. The critical point in his treatment is surely the finding of chd ½ from chd 1½ and, as we have seen, his treatment of this point is neither simple nor accurate. Since this is so, it does not seem correct to describe his overall treatment as beautiful or elegant.

In summary, Ptolemy gives a table of chords with a precision of three sexagesimal positions, which gives about twice the precision of five decimal positions. However, there are a number of errors of one unit in the third position, and the overall accuracy of Ptolemy's table is about equivalent to that of five decimal positions.

Note added in proof: Additional research done since I wrote this appendix shows that the value of 70;32,04 for chd 72° in Table C.1 is a scribal error. Ptolemy's original value, which is also the correct one, is 70;32,03. I am writing up this research for publication elsewhere.

TABLE C.1

CORRECTIONS TO PTOLEMY'S TABLE OF CHORDS

x, *degrees*	Chord *x*	Ptolemy's Value[a]	*x*, *degrees*	Chord *x*	Ptolemy's Value[a]
2	2 05 39.44	40	82½	79 07 17.39	18
3½	3 39 52.64	52	84	80 17 44.42	45
4	4 11 16.58	16	84½	80 41 02.46	03
10	10 27 31.28	32	86	81 50 23.29	24
14	14 37 27.56	27	88	83 21 32.42	33
18	18 46 19.69	19	88½	83 44 05.48	06
20½	21 21 11.61	11	91½	85 57 22.44	23
26	26 59 38.86	38	92½	86 41 01.23	02
26½	27 30 14.57	14	93½	87 24 16.26	17
30½	31 33 49.48	50	98½	90 54 28.07	29
49	49 45 47.48	48	99	91 14 55.38	56
53	53 32 37.46	38	99½	91 35 16.43	17
55	55 24 35.40	36	100	91 55 31.20	32
55½	55 52 25.47	26	102½	93 35 10.10	11
58½	58 38 04.38	05	107	96 27 46.16	47
59½	59 32 45.53	45	108	97 04 55.34	56
60½	60 27 10.36	11	110½	98 35 51.48	52
62	61 48 16.45	17	112	99 29 04.23	05
63	62 41 59.38	60	114½	100 55 28.85	28
63½	63 08 44.41	45	115½	101 29 14.42	15
64½	64 02 01.47	02	116½	102 02 32.16	33
65	64 28 33.43	34	118	102 51 36.27	37
67½	66 40 06.34	07	121½	104 41 58.28	59
68	67 06 11.33	12	123	105 27 28.99	30
69	67 58 07.49	08	123½	105 42 24.80	26
72	70 32 03.23	04	124	105 57 13.36	14
73	71 22 43.44	44	125½	106 40 55.40	56
77	74 42 06.32	07	127½	107 37 29.02	30
78	75 31 06.41	07	128	107 51 19.03	20
80	77 08 04.25	05	131	109 11 43.27	44

[a] This column gives Ptolemy's third sexagesimal position only. His table is correct in the first and second positions.

TABLE C.1 (continued)

x, *degrees*	Chord *x*	Ptolemy's Value[a]	*x*, *degrees*	Chord *x*	Ptolemy's Value[a]
131½	109 24 41.20	42	162	118 31 21.36	22
134	110 27 38.10	39	162½	118 36 12.17	13
134½	110 39 50.82	52	163½	118 45 29.40	30
135	110 51 55.96	57	164½	118 54 14.07	15
135½	111 03 53.50	54	165	118 58 24.18	25
136	111 15 43.42	44	167	119 13 43.04	44
137	111 39 00.39	01	167½	119 17 12.34	13
137½	111 50 27.40	28	168	119 20 33.46	34
138½	112 12 58.41	59	168½	119 23 46.40	47
139	112 24 02.39	03	169	119 26 51.16	52
139½	112 34 58.66	60	169½	119 29 47.73	49
140	112 45 47.21	48	170	119 32 36.11	37
140½	112 56 28.04	29	170½	119 35 16.30	17
141	113 07 01.12	02	171	119 37 48.29	49
144	114 07 36.42	37	171½	119 40 12.08	13
148	115 21 05.05	06	173	119 46 34.23	35
149	115 38 08.36	09	173½	119 48 25.20	26
149½	115 46 28.12	29	175	119 53 08.83	10
157	117 35 27.47	28	176	119 55 36.84	38
157½	117 41 39.24	40	177½	119 58 17.20	18
159	117 59 26.12	27	178	119 58 54.20	55
159½	118 05 05.58	07	178½	119 59 22.99	24
160½	118 16 00.22	01			
161	118 21 15.38	16			
161½	118 26 22.43	23			

[a]This column gives Ptolemy's third sexagesimal position only. His table is correct in the first and second positions.

TABLE C.2

A LIST OF ANGLES
FOR WHICH THERE ARE DISCREPANT CHORD VALUES[a]

Angle, *degrees*	Angle, *degrees*
9	89
20½	89½
31	97
39	102½
51	118½
64	141½
72	143
77½	166½
85	172
88½	172½
	174

[a]That is, for which there are discrepancies between the texts of Halma and Heiberg.

APPENDIX D

PTOLEMY'S TABLE OF THE LATITUDE AS A FUNCTION OF THE LONGEST DAY OF THE YEAR

In Chapter II.6 of the *Syntaxis*, Ptolemy gives a table of the latitude of a spot as a function of the length of the longest day of the year at that spot; for brevity I shall call this his latitude table. He also gives some other quantities that are functions of the latitude, but I shall not take them up here. In calculating his latitude table, Ptolemy takes the obliquity of the ecliptic to be 23;51,20 degrees.

In calculating this table, Ptolemy takes the sun to be a point, he takes the earth to be a perfect sphere, and he neglects parallax and refraction. The problem of relating the latitude to the longest day, under these idealizing assumptions, can be studied with the aid of Figure D.1. I believe that this figure is easier to follow than the one Ptolemy draws, but it is based upon his figure in one important respect: After considerable experimentation, I decided that it is best to follow him and to take the plane of the figure to be the observer's meridian plane rather than his horizon.

In Figure D.1, then, the circle *NPZBC* represents the circle that the observer's meridian cuts on the celestial sphere, and the space within the circle represents the half of the celestial sphere that lies east of the meridian. All lines in the figure represent arcs of great circles drawn on the celestial sphere. The arc *NSE* is the quarter circle cut on the celestial sphere by the northeastern quadrant of the horizon. Point *E* is the point on the horizon that is due east, while point *N* is due north. *P* is the North Pole and arc *BEHC* is the equator.

When the sun rises, it is by definition on the eastern horizon, so it is at some point *S* on the horizon arc *NSE*. The plane through the sun and through the axis of rotation necessarily passes through points *P* and *S*,

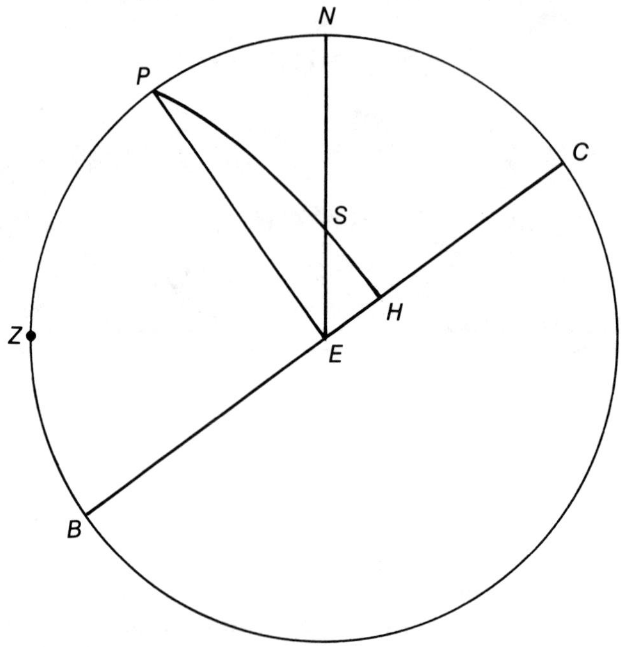

Fig. D.1—The relation between the latitude, the time of year, and the length of the day. The circle *NPZBC* is the circle cut on the celestial sphere by the observer's meridian, and the space within this circle is the half of the celestial sphere that lies east of the meridian. Point *P* is the North Pole, so that arc *NP* is the latitude. Arc *BH* is half of the length of the day, after conversion of time to angle, and arc *SH* is the declination of the sun. The angle at *H* is a right angle.

and arc *PSH* is the quarter circle cut by this plane on the celestial sphere. The part *SH* of this arc is the declination δ of the sun, which is assumed to be constant during the day in question; this is not necessarily the longest day as the figure is drawn.

Let L denote the length of the day in hours. Then $15L$ is the number of degrees by which the earth (or the sun in the geocentric picture) rotates between sunrise and sunset.[1] Half this amount, or $15L/2$, is the angle between the meridian plane and the plane containing the sun at sunrise; this is the plane through the arc *PSH*. The arc *BH* of the equator is the

[1]Note that we neglect the daily motion of the sun in this derivation.

arc subtended by the angle $15L/2$. We are also interested in the arc *EH*, which I shall denote by λ. Thus

$$\lambda = \text{arc}\, EH = (15L/2) - 90^\circ. \quad \text{(D.1)}$$

PE is part of the great circle cut by the plane through the pole *P* and the point *E* that is due east. Since the arc *NP* in the meridian circle is the latitude ϕ of the observer, the angle *PEN* between arcs *PE* and *NSE* also equals ϕ.

Now consider the spherical triangle *SEH*. Angle *H* is a right angle by construction, arc *SH* equals the declination δ of the sun, and arc *EH* is the arc λ defined by Equation D.1. Since λ and δ are the given quantities in the problem, the triangle is completely determined, and we can solve for the angle *E*. Since *E* is obviously the complement of ϕ, we have now calculated the latitude ϕ from the length of the day.

Before I discuss the triangle *SEH* further, I want to introduce another quantity. The arc *NS* is the azimuth of the rising point of the sun, measured along the horizon from the north point *N* to the rising point *S*. I shall denote this azimuth by *A*:

$$A = \text{arc}\, NS. \quad \text{(D.2)}$$

Elementary spherical trigonometry supplies a number of relations that connect the elements of the triangle *SEH*. Among them are:

$$\begin{aligned} \sin\lambda &= \tan\delta\tan\phi, \\ \sin A &= \cos\delta\cos\lambda, \\ \sin\phi &= \tan A\tan\lambda. \end{aligned} \quad \text{(D.3)}$$

Ptolemy of course did not know these and other relations in forms like the modern Equations D.3, but he did know certain methods of calculation that are equivalent to some of them. Thus, for simplicity, I shall speak in terms of the modern equivalents to the methods that he did know.

Since δ and λ are the "givens" in the problem and ϕ is the quantity to be solved for, it is obvious to us that we should use the first of Equations D.3. However, *Pedersen* [1974, pp. 75-78] has studied the methods of solving spherical triangles that Ptolemy uses in the *Syntaxis*, and the first equation does not seem to be among them. This suggests that the first equation was unknown to Greek astronomers.

Whether he did not know the first of Equations D.3, or whether he did not think to use it, Ptolemy solved for ϕ indirectly. He first used δ and λ to find the azimuth *A* from the second of Equations D.3. Then, with *A* known, he used the last equation to find ϕ.

For the sake of emphasis, I repeat that Ptolemy did not know Equations D.3 or other equations like them. When I say that he used one of these equations, I am using the equation as a succinct identification of a lengthy process that produces the same answer, a process in which Ptolemy could use no trigonometric function except the chord.

Figure D.1 and Equations D.3 apply for any declination δ of the sun and hence for any length of the day. On the longest day, the sun is farthest north. Thus, when we specialize to the longest day, we must replace δ by ϵ, the obliquity of the ecliptic, throughout.

Now let us take up Ptolemy's table. In Table D.1, the first column gives the lengths of the longest day that Ptolemy considers. The length goes by increments of ¼ hour from 12 hours through 18 hours. It then goes by increments of ½ hour through 20 hours, and finally by increments of 1 hour through 24 hours. The column labelled ϕ_P in Table D.1 gives the latitude that Ptolemy tabulates. Ptolemy gives the latitude to a precision of 1 minute of arc through most of the table. For lengths of the day longer than 16½ hours, however, he drops the precision to 10′ or even to 30′. There is one exception: Where the longest day is 24 hours, we know without calculation that the latitude is the complement of the obliquity, and Ptolemy gives this value to the second of arc.

Where the longest day is 16½ hours, the Greek texts of both Halma and Heiberg give the latitude as the equivalent of 51° 40′. This value is clearly out of line with the others, as Manitius notes in his translation of Heiberg's text. I have no doubt that Ptolemy meant to write the equivalent of 51° 30′, and I shall assume this value in the remaining discussion.

Ptolemy does not actually present these results in tabular form, as he does with his table of chords that we discussed in Appendix C. Instead, he devotes a short paragraph to each length of the day that he considers. In this paragraph, he gives the relevant length of the longest day and the latitude that he calculates from it. He then gives the length of the shadow cast at noon by a vertical pole (a gnomon) of length 60; he gives this shadow length at the equinoxes and at both solstices. He may also give other interesting information in the paragraph, such as a city or region having the latitude in question. I shall not consider any information here except the length of day and the latitude.

Ptolemy also gives some information about latitudes north of the Arctic Circle, where the length of uninterrupted sunlight can range up to 6 months. I shall not take the space to describe this information.

In the paragraph relating to a particular length of day, Ptolemy does not give the azimuth A, which was defined in Equation D.2. However, we know that he uses A as an intermediary result in finding ϕ because he describes his method of calculating ϕ in detail in Chapters II.2 and II.3 of the *Syntaxis*. He devotes Chapter II.2 to solving the second of Equa-

TABLE D.1

LATITUDE AS A FUNCTION OF THE LONGEST DAY OF THE YEAR[a]

Length of Longest Day, *hours*	ϕ_P[b] ° ′	A_M ° ′ ″	ϕ_M[c] ° ′ ″	Apparent A ° ′ ″
12	0 00	66 08 40	0 00 00	66 08 40
12¼	4 15	66 04 31	4 13 54	66 10 01
12½	8 25	65 52 07	8 24 47	65 52 39
12¾	12 30	65 31 41	12 29 52	65 31 55
13	16 27	65 03 33	16 26 42	65 03 57
13¼	20 14	64 28 08	20 13 19	64 28 51
13½	23 51	63 45 57	23 48 20	63 48 20
13¾	27 12	62 57 30	27 10 50	62 58 25
14	30 22	62 03 21	30 20 23	62 04 30
14¼	33 18	61 04 02	33 16 57	61 04 42
14½	36	60 00 03	36 00 48	59 59 35
14¾	38 35	58 51 53	38 32 22	58 53 21
15	40 56	57 40 00	40 52 21	57 41 54
15¼	43 04	56 24 47	43 01 25	56 26 04
15½	45 01	55 06 36	45 00 19	55 06 55
15¾	46 51	53 45 46	46 49 47	53 46 19
16	48 32	52 22 35	48 30 34	52 23 12
16¼	50 04	50 57 19	50 03 25	50 57 33
16½	51 40	49 30 09	51 28 53	49 30 35
16¾	52 50	48 01 20	52 47 41	
17	54 30	46 31 00	54 00 17	
17¼	55	44 59 20	55 07 15	
17½	56	43 26 28	56 09 04	
17¾	57	41 52 31	57 06 06	
18	58	40 17 35	57 58 44	

[a] All the calculations in this table are based upon the value 23°51′20″ for the obliquity of the ecliptic.

[b] The latitude tabulated by Ptolemy.

[c] The latitude calculated by modern methods.

TABLE D.1 (continued)

Length of Longest Day, *hours*	ϕ_P[b] ° ′	A_M ° ′ ″	ϕ_M[c] ° ′ ″	Apparent A ° ′ ″
18½	59½	37 05 10	59 32 12	
19	61	33 49 54	60 51 55	
19½	62	30 32 16	61 59 38	
20	63	27 12 43	62 57 00	
21	64½	20 29 12	64 25 19	
22	65½	13 41 32	65 24 02	
23	66	6 51 22	65 57 48	
24	66 08[d]	0 00 00	66 08 40	

[b]The latitude tabulated by Ptolemy.
[c]The latitude calculated by modern methods.
[d]More specifically, Ptolemy gives 66°08′40″ for this latitude. This value is the complement of the obliquity.

tions D.3 for A and he devotes Chapter II.3 to solving the third equation for ϕ. In his descriptions, he uses a day of 14½ hours as an example.

In calculating A by his method, Ptolemy needs chd $2(90° - \lambda)$ and chd $2(90° - \epsilon)$, and he finds them by interpolating in his table. I have performed these interpolations and get the same values that he does. He then combines the values in a way that yields chd $2A$, and he states that the result is 103;55,23. This is exactly the value of chd 120° that appears in his table, so he concludes that the azimuth A is exactly 60° at the latitude where the longest day is 14½ hours. In other words, at the summer solstice, the sun rises 30° north of due east at this latitude.

In finding ϕ from A and λ, Ptolemy needs chd $2(90° - \lambda)$, which he has already used, chd 2λ, which he has not yet used, chd $2A$, which he has just found, and chd $2(90° - A)$. Since A is exactly 60°, the latter value is exactly 60. Since $\lambda = 18;45$ degrees, $2\lambda = 37½$, which is one of the entries in his table of chords. Thus chd $2\lambda = 38;34,22$ without the need to interpolate. From these values Ptolemy finds chd $2\phi = 70;33$, whence $2\phi = 72;1$ degrees and ϕ is almost exactly 36°.

Incidentally, we can see why the latitude 36° was so highly regarded by Greek astronomers, aside from the fact that it was nominally the latitude of Hipparchus's observatory in Rhodes.[2] It was generally taken

[2]I find it personally satisfying that a street in the old part of the city of Rhodes is named for Hipparchus.

as the latitude where the longest day had an exact value, namely 14½ hours, and where the sun rose on the longest day exactly 30° north of east. In addition, these same values are produced by the numerological method of finding the obliquity that *Moesgaard* [1980] has suggested; see Section II.7.

Several comments are needed on Ptolemy's example. It seems miraculous that chd $2A$ should come out to be exactly the value that is tabulated for 120°, but this is another miracle of Ptolemy's making. When the arithemtic is done accurately, we find chd $2A$ = 103;55,26, which leads to A = 60;0,3 degrees. Perhaps we should not criticize Ptolemy for altering this to 60° without comment, although a modern mind would prefer him to point out the approximation.

If Ptolemy had used the value of chd $2A$ that actually came from the solution, he would have had chd $2(90° - A)$ = 59;59,55 rather than 60. This would have given him chd 2ϕ = 70;33,23 and ϕ = 36;0,47. Using the value A = 60°, he should have gotten chd 2ϕ = 70;33,15 and ϕ = 36;0,42. Either way, he had to neglect more than 40″ in order to round to 36° exactly.

In studying Ptolemy's other values of the latitude in Table D.1, I think we can use modern methods of solution rather than Ptolemy's, provided that we observe suitable precautions. There are two reasons for this:

a. The method of chords is as valid as modern methods; it is just less convenient. If we solve the problem by modern methods, we should get exactly the same answers as Ptolemy if all the calculations, both his and ours, are done accurately.
b. The limiting factor in Ptolemy's accuracy is his table of chords. We saw in Appendix C that angles found from this table are accurate to about 1 second of arc. However, we shall find errors in his latitudes that are measured in minutes, not seconds. Thus his errors did not come from his chord table, and thus we may treat his chord table as exact in this appendix. This further means that we can use modern functions in studying his accuracy.

However, we may not proceed to calculate ϕ directly by using the first of Equations D.3. Ptolemy first found chd $2A$ by a sequence of multiplications and divisions. He then found $2A$ by interpolation in his chord table, and he next found chd $2(90° - A)$ by another interpolation.[3] These facts mean that his method does not yield exactly the same results that we would get by using the direct solution. In other words, we must first find A by using the second of Equations D.3, and we must

[3]Ptolemy did not have to interpolate in his example, but he would have had to in most cases.

round this value to the second of arc; this is the level of rounding that Ptolemy should have used. We then enter the last of Equations D.3 with this rounded value of A and solve for ϕ, which we also round to the second of arc.

The test of Ptolemy's table is shown in the third and fourth columns of Table D.1. The column labelled A_M is the value of the azimuth A, found in the way just described, and the column labelled ϕ_M is the corresponding value of ϕ. As I noted above, the difference between Ptolemy's ϕ and the modern ϕ is measured in minutes of arc, well beyond any reasonable error in calculation in Ptolemy's time, and we naturally seek an explanation of his errors.

One possible explanation comes from the fact that Ptolemy, in his example, rounded the value of A instead of using it to the nearest second of arc. In order to see whether rounding the intermediary quantity A could account for the errors, I have calculated the fifth column in Table D.1, the column labelled Apparent A. In order to calculate this column, I substituted the values of λ and ϕ from Ptolemy's table into the last of Equations D.3 and solved for A. I calculated this apparent A for lengths of the day only through 16½ hours, since Ptolemy is obviously approximating ϕ beyond this point.

In using the apparent A's, we must remember that the value of ϕ_P itself has been rounded to the minute of arc in most of the table, so the apparent A is not necessarily the A that Ptolemy used in finding ϕ_P. Even when we keep this in mind, however, I do not see any reasonable way in which the values of A_M could have been rounded into values that would yield the column ϕ_P. There are several reasons for this conclusion, but the most important one concerns the systematic nature of the errors.

In considering this point, we should observe that the line for a day of 14½ hours is an anomaly. This is the example that Ptolemy used, where we saw him deliberately introduce errors in order to obtain $A = 60°$ and $\phi = 36°$, exactly. Aside from this line, however, there is a remarkable fact: For lengths of the day through 16½ hours, the value of A_M is smaller than the apparent A in every instance, and Ptolemy's ϕ_P is larger than the correct ϕ_M in every instance. It does not seem possible that this systematic effect resulted from rounding, which is inherently random in its direction.

Now we ask whether Ptolemy might have introduced a systematic error either by truncating, which amounts to rounding down on all occasions, or by always rounding up. If Ptolemy did either here, he departed from his practice in the rest of the *Syntaxis*. His general practice was to round to a nearby value that he regarded as convenient, regardless of the direction of the rounding.

The explanation of Ptolemy's error, I believe, is one that is suggested by the fact that led me to investigate his latitude table in the first place: Ptolemy asserts that the length of the longest day on the Tropic of Cancer (that is, where the latitude equals the obliquity) is exactly 13½ hours, and he also asserts that the obliquity is 23° 51′ 20″. As we saw in Section II.7, these statements are not compatible. If the longest day on the Tropic of Cancer is 13½ hours, the obliquity is 23° 49′ 50″ rather than 23° 51′ 20″.

Suppose either that Ptolemy made this error accidentally, or that some earlier astronomer made it and that Ptolemy repeated it without checking. This fact should not affect the accuracy of the rest of Ptolemy's latitude table. That is, if we accept the value 23° 51′ 20″ for the obliquity, as Ptolemy explicitly did, and if we calculate the value of ϕ by using this obliquity, we should obtain Ptolemy's values within reasonable error. As the columns ϕ_P and ϕ_M in Table D.1 show, we do not. Thus the error in relating the obliquity and the longest day is not an accidental and isolated error.

Instead, as Table D.1 shows us, there is a systematic error in relating the latitude to the length of the longest day. The simplest explanation of this fact, it seems to me, is that the latitude table, including the value of the obliquity which makes the longest day exactly 13½ hours on the Tropic, was calculated at a time when the errors in trigonometric tables were measured in minutes rather than seconds.[4]

Eratosthenes is specifically associated with the value of 23° 51′ 20″ for the obliquity, and he is specifically associated with the statement that Aswan (Syene) lies on the Tropic of Cancer. Even in his time, as I have shown [*Newton*, 1980], it is almost impossible that a measurement of the latitude at Aswan would have put it either on the Tropic or at the latitude 23° 51′ 20″, and it is almost impossible that a measurement of the obliquity would have yielded the value 23° 51′ 20″. The only plausible explanation of the obliquity, it seems to me, is that it was calculated from the day length of 13½ hours, and that an error was made in the calculation.

As I said in Section II.7, attributing the origin of the obliquity to an error in calculation would be implausible except for the fact that Ptolemy explicitly reproduces the error in his table. Thus we know that the error actually occurred. Also we now know that the error was not an isolated one, and an error in calculating the obliquity has actually become plausible. Still further, since Eratosthenes is so intimately associated with the erroneous value of the obliquity, we conclude that the error originated with him or in his time.

[4]The lengths of the shadow cast by a gnomon that appear in Ptolemy's table have been rounded to a multiple of 5′. It does not seem likely that we can learn much about errors in trigonometric calculations from studying such rounded numbers, and I have not considered them in this appendix.

In summary of the argument up to this point, the latitude table was calculated at a time when the errors in trigonometric tables were minutes of arc rather than seconds, and thus the table was not calculated by Ptolemy.[5] The best evidence we have is that the table originated in the time of Eratosthenes, and it is quite possible that he was the author of it.

Incidentally, Ptolemy's table of chords is given to three sexagesimal positions, and it gives an accuracy of about 1″ when we use it to find an angle. The table of chords used to calculate the latitude table gave an accuracy of about 1′, and thus it is plausible that Eratosthenes or his contemporaries had a table of chords given to two sexagesimal positions.

As I mentioned in Section II.7, *Moesgaard* [1980] has an interesting and ingenious suggestion about the origin of the value 23° 51′ 20″. His explanation requires the value to be obtained from calculations that were accurate to the second of arc. However, as Table D.1 clearly shows, the value of 23° 51′ 20″ was already in use at a time when trigonometric calculations could be made with an accuracy of only a minute or less. Thus Moesgaard's explanation seems unlikely to me, although it is certainly a remarkable fact that it yields the specific value 23° 51′ 20″. The most likely explanation, as I have already said, is that the value arose from errors in an ancient table of the trigonometric functions.

Before concluding this appendix, it is useful to return briefly to Eratosthenes' determination of the size of the earth. In finding the size, he assumed that Aswan and Alexandria are on the same meridian and that they are separated in latitude by 1/50 of a circle. We know that he took the latitude of Aswan to be 23;51 degrees, dropping the seconds position for brevity, and it now seems safe to conclude that he also thought that Aswan was on the tropic and that the longest day there was 13½ hours.

We do not know the latitude that Eratosthenes took for Alexandria, but we do know that Ptolemy used 30;58 degrees. We also know [*Newton*, 1980] that 30;58 = $\tan^{-1}$ 0.6, to the nearest minute of arc, and we know that what we call the tangent of the latitude was important in Greek astronomy and geography. Hence I suggested that Eratosthenes took the tangent to be exactly 0.6 and that he calculated the value 30;58 from this. He then subtracted 23;51 degrees from 30;58, getting 7;7 degrees, which he rounded to be 1/50 of a circle (7;12 degrees).

This appendix shows that we need to revise this conclusion slightly. If Eratosthenes made the mistake of finding 23;51,20 rather than 23;49,50 for the obliquity, he would not have found 30;58 for $\tan^{-1}$ 0.6.

[5]The latitude of 36° where the longest day is 14½ hours is an exception to this statement. Ptolemy explicitly calculated this value as an example, but he obviously altered his calculations in order to get the value 36°. However, the attraction of the value 36° was probably a strong one for all Greek astronomers, and it is plausible that this value already appeared in the table that Ptolemy took over without acknowledgment.

Now finding both values involves the Greek equivalent of finding the arctangent, and Table D.1 is compatible with the use of tables which give too large a value for the arctangent by a few minutes of arc. Thus, if Eratosthenes used $\tan^{-1} 0.6$ for the latitude of Alexandria, it is likely that he would have found 30;59 or 31;00 rather than 30;58.

Then, when he found the arc between Aswan and Alexandria, he would have found 7;8 or 7;9 degrees. Either value is easier to round to 1/50 of a circle than is 7;7 degrees.

Thus everything we know about Eratosthenes' determination is consistent with the idea that he took the latitude of Alexandria to be $\tan^{-1} 0.6$ exactly, and that he found the latitude by using trigonometric tables that are in error in the minutes position. His value for the latitude has been lost and we do not know exactly what it was. However, since the value 30;58 degrees is correct to the minute, this value must have been calculated by someone between Eratosthenes and Ptolemy.

If the conclusions of this appendix are correct, we can put the appearance of trigonometric tables within a narrow span of time. They were not available to Aristarchus when he estimated the sizes of the sun and moon, but they were available in the time of Eratosthenes about a generation later.

REFERENCES

American Ephemeris and Nautical Almanac, U. S. Government Printing Office, Washington, published annually.

Aristarchus, *On the Sizes of the Sun and Moon*, ca. −280. There is an edition, with a parallel translation into English, by Sir Thomas Heath in *Aristarchus of Samos*, Oxford University Press, Oxford, 1913.

Britton, J. P., On the quality of solar and lunar observations and parameters in Ptolemy's *Almagest*, a dissertation submitted to Yale University, 1967.

Britton, J. P., Ptolemy's determination of the obliquity of the ecliptic, *Centaurus,* **14**, pp. 29-41, 1969.

CCP = *Newton* [1977].

Censorinus, *Liber de Die Natali*, 238. There is an edition, with a parallel translation into French, by Desiré Nisard in *Celse, Vitruve, Censorin, et Frontin*, Dubochet et Cie., Paris, 1846.

Copernicus, Nicolaus, *De Revolutionibus Orbium Caelestium Libri Sex*, 1543. There is an edition, including a facsimile of Copernicus's holograph, by Fritz Kubach, Verlag von R. Oldenbourg, Munich, in two volumes: volume 1, facsimile of the holograph, 1944; volume 2, a critical edition of the text, 1949.

Delambre, J. B. J., *Histoire de l'Astronomie du Moyen Âge,* Chez Mme. Veuve Courcier, Paris, 1819.

Dreyer, J. L. E., *History of the Planetary Systems from Thales to Kepler*, 1905. This has been republished under the title *A History of Astronomy from Thales to Kepler*, Dover Publications, New York, 1953. Page numbers are from this republication.

Dreyer, J. L. E., *Tychonis Brahe Dani Opera Omnia*, in 13 volumes, Libraria Gyldendaliana, Copenhagen, 1926. (There are 15 volumes in the complete series, but the last two have other editors and I do not cite them here.)

Explanatory Supplement to The Astronomical Ephemeris and The American Ephemeris and Nautical Almanac, H. M. Stationery Office, London, 1961.

Geminus, *Eisagoge Eis ta Phainomena*, ca. −100. There is an edition under the title *Elementa Astronomiae*, with a parallel translation into German, by K. Manitius, B. G. Teubner, Leipzig, 1898. The date −100 is a guess. Geminus could not have been much earlier than this, but he could have been as much as two centuries later.

Gingerich, O., Was Ptolemy a fraud?, *Quarterly Journal of the Royal Astronomical Society*, **21**, pp. 253-266, 1980.

Gingerich, O., Ptolemy revisited: a reply to R. R. Newton, *Quarterly Journal of the Royal Astronomical Society*, **22**, pp. 40-44, 1981.

Goldstein, B. R., The Arabic version of Ptolemy's planetary hypotheses, *Transactions of the American Philosophical Society*, **57**, Part 4, 1967.

Goldstein, B. R., Review of *The Crime of Claudius Ptolemy, Science*, **199**, p. 872, 1978.

Harrington, J. P., Review of *The Crime of Claudius Ptolemy, Archaeoastronomy Bulletin*, **1**, pp. 9-12, 1978.

Hartner, W., The role of observations in ancient and medieval astronomy, *Journal for the History of Astronomy,* **8**, pp. 1-11, 1977.

Heath, Sir Thomas, *Aristarchus of Samos*, Oxford University Press, Oxford, 1913.

Herodotus, *History*, ca. −446. There is a translation into English by George Rawlinson, first published in 1858 and reprinted by Tudor Publishing Co., New York, 1947.

Hill, G. W., Tables of Jupiter, constructed in accordance with the methods of Hansen, *Astronomical Papers Prepared for the Use of The American Ephemeris and Nautical Almanac,* **VII**, Part 1, U. S. Government Printing Office, Washington, 1895.

Hipparchus, *In Arati et Eudoxi Phaenomena Commentariorum Libri Tres*, ca. −135. There is an edition, with a parallel translation into German, by K. Manitius, B. G. Teubner, Leipzig, 1894.

ibn Yunis, *Az-Zij al-Kabir al-Hakimi*, 1008. There is an edition of part of this work, with a parallel translation into French, by J. J. A. Caussin de Perceval, under the title *Le Livre de la Grande Table Hakémite*, Imprimerie de la République, Paris, 1804.

Menelaos, *Sphaerica*, ca. 100. Abu Nasr Mansur bin Ali bin Iraq prepared an Arabic version around the year 1000. There is a translation of this version into German by Max Krause under the title *Die Sphärik von Menelaos aus Alexandrien*, Weidmannsche Buchhandlung, Berlin, 1936. The work is lost in the original language.

Moesgaard, K. P., Hipparchus's solar theory derived from lunar eclipse observations, a paper presented at a meeting of the History of Astronomy Commission of the International Astronomical Union, Grenoble, August 1976. This paper was summarized by M. A. Hoskin in *Journal for the History of Astronomy*, **7**, pp. 216-217, 1976. See also Moesgaard's further comments following the paper cited as [*Newton*, 1977a] that did not receive a separate title.

Moesgaard, K. P., The full moon serpent. A foundation stone of ancient astronomy?, *Centaurus,* **24**, pp. 51-96, 1980.

Neugebauer, O., *Astronomical Cuneiform Texts*, in 3 volumes, Lund Humphries, London, 1955.

Neugebauer, O., *The Exact Sciences in Antiquity, Second Edition*, Brown University Press, Providence, Rhode Island, 1957.

Neugebauer, O., *A History of Ancient Mathematical Astronomy*, Springer-Verlag, New York, 1975.

Newcomb, S., Tables of the motion of the earth on its axis and around the sun, *Astronomical Papers Prepared for the Use of The American Ephemeris and Nautical Almanac*, **VI**, Part 1, U.S. Government Printing Office, Washington, 1895.

Newcomb, S. Tables of Mercury, *Astronomical Papers Prepared for the Use of The American Ephemeris and Nautical Almanac,* **VI**, Part 2, U.S. Government Printing Office, Washington, 1895a.

Newton, R. R., *Ancient Astronomical Observations and the Accelerations of the Earth and Moon*, Johns Hopkins University Press, Baltimore and London, 1970.

Newton, R. R., The obliquity of the ecliptic two millenia ago, *Monthly Notices of the Royal Astronomical Society,* **169**, pp. 331-342, 1974.

Newton, R. R., *Ancient Planetary Observations and the Validity of Ephemeris Time*, Johns Hopkins University Press, Baltimore and London, 1976.

Newton, R. R., *The Crime of Claudius Ptolemy*, Johns Hopkins University Press, Baltimore and London, 1977.

Newton, R. R., Remarks on "Hipparchus's Solar Theory Derived from Lunar Eclipses" by K. P. Moesgaard, *Journal for the History of Astronomy,* **8**, pp. 200-203, 1977a. Note especially Moesgaard's reply to my remarks that immediately follows my paper, but without a separate title.

Newton, R. R., On a pedagogical motivation for Ptolemy's fabrication of data, *Archaeoastronomy Bulletin,* **2**, pp. 7-9, 1978.

Newton, R. R., On the fractions of degrees in an ancient star catalogue, *Quarterly Journal of the Royal Astronomical Society*, **20**, pp. 383-394, 1979.

Newton, R. R., The sources of Eratosthenes' measurement of the earth, *Quarterly Journal of the Royal Astronomical Society,* **21**, pp. 379-387, 1980.

Newton, R. R., Comments on "Was Ptolemy a Fraud?" by Owen Gingerich, *Quarterly Journal of the Royal Astronomical Society,* **21**, pp. 388-399, 1980a.

Oppolzer, T. R. von, *Canon der Finsternisse*, Kaiserlich-Königlichen Hof- und Staatsdruckerei, Vienna, 1887. There is a reprint, with the explanation of the tables translated into English by O. Gingerich, by Dover Publishing Co, New York, 1962.

Pannekoek, A., Ptolemy's precession, *Vistas in Astronomy,* **1**, pp. 60-66, 1955.

Pannekoek, A., *A History of Astronomy*, Interscience Publishers, New York, 1961.

Pedersen, Olaf, *A Survey of the Almagest*, Odense University Press, Odense, Denmark, 1974.

Peters, C. H. F., and Knobel, E. B., *Ptolemy's Catalogue of Stars*, Carnegie Institute of Washington, Washington, 1915.

Plato, *Phaedo*, ca. −380. There is a translation into English by B. Jowett which is reprinted in *Harvard Classics,* volume 2, pp. 45-113, P. F. Collier & Son, New York, 1909.

Ptolemy, C., *'E Mathematike Syntaxis*, ca. 142. There is an edition by J. L. Heiberg in *C. Ptolemaei Opera Quae Exstant Omnia*, B. G. Teubner, Leipzig, 1898. There is a translation of this edition into German by K. Manitius, B. G. Teubner, Leipzig, 1913. There is also an edition with a parallel translation into French by N. B. Halma, Henri Grand Libraire, Paris, 1813. There is also a translation of Heiberg's edition into English by R. C. Taliaferro in *Great Books of the Western World,* volume 16, Encyclopaedia Britannica, Inc., Chicago, 1952. Quotations are from this translation unless they are otherwise attributed.

Rawlins, Dennis, The unreality of extant ancient mathematical geography, in preparation.

Rawlins, Dennis, An investigation of the ancient star catalogue, *Publications of the Astronomical Society of the Pacific,* **94**, pp. 359-373, 1982.

Scientific American, Science and the Citizen, **240**, p. 92, March, 1979.

Swerdlow, N. M., Ptolemy on trial, *The American Scholar,* **48**, pp. 523-531, 1979.

Syntaxis = *Ptolemy* [ca. 142].

Tannery, Paul, *Recherches sur l'Histoire de l'Astronomie Ancienne*, Gauthier-Villars et Fils, Paris, 1893.

Times Atlas of the World, Mid-Century Edition, in 5 volumes, The Times Office, London, 1955.

Toomer, G. J., Apollonius of Perga, *Dictionary of Scientific Biography,* volume I, pp. 179-193, Charles Scribner's Sons, New York, 1970.

Toomer, G. J., Ptolemy, *Dictionary of Scientific Biography,* volume XI, pp. 186-206, Charles Scribner's Sons, New York, 1975.

Wells, H. G., *The Outline of History, Third Edition*, The Macmillan Company, New York, 1922.

INDEX

G

H

I

J

K

L

M

N

About the Author

Robert R. Newton is the Supervisor of the Space Sciences Branch of The Johns Hopkins University Applied Physics Laboratory. Dr. Newton received his Ph.D. from Ohio State University in 1946, worked for the Bell Telephone Laboratories for two years, and taught at the University of Tennessee and at Tulane University before joining the staff of the Applied Physics Laboratory. His early work at APL involved the study of the flight mechanics of satellites and space probes. More recently, his primary area of interest has been the use of ancient and medieval astronomical observations to study the time variation of nongravitational forces in the solar system. This work has resulted in numerous articles and several books (listed on page ii).